Justification of the Inventive Activity

Thomas Heinz Meitinger

Justification of the Inventive Activity

The proceeding before the DPMA, the EPO, and the Federal Patent Court

Springer

Thomas Heinz Meitinger
Meitinger Patentanwalts GmbH
Munich, Germany

ISBN 978-3-662-72734-8 ISBN 978-3-662-72735-5 (eBook)
https://doi.org/10.1007/978-3-662-72735-5

Translation from the German language edition: "Begründung der erfinderischen Tätigkeit" by Thomas Heinz Meitinger, © Der/die Herausgeber bzw. der/die Autor(en), exklusiv lizenziert an Springer-Verlag GmbH, DE, ein Teil von Springer Nature 2025. Published by Springer Berlin Heidelberg. All Rights Reserved.

This book is a translation of the original German edition "Begründung der erfinderischen Tätigkeit" by Thomas Heinz Meitinger, published by Springer-Verlag GmbH, DE in 2025. The translation was done with the help of an artificial intelligence machine translation tool. A subsequent human revision was done primarily in terms of content, so that the book will read stylistically differently from a conventional translation. Springer Nature works continuously to further the development of tools for the production of books and on the related technologies to support the authors.

This Springer imprint is published by the registered company Springer-Verlag GmbH, DE, part of Springer Nature.
The registered company address is: Heidelberger Platz 3, 14197 Berlin, Germany

Preface

A patent can be a very important asset for a company. However, the path to obtaining one is challenging. Only inventions that are not yet found anywhere else and that also demonstrate an inventive step are granted a patent. The key criteria for patent granting are therefore novelty and inventive step.

The justification of inventive step regularly determines the outcome of grant proceedings before the patent offices, the German Patent and Trademark Office (DPMA) and the European Patent Office (EPO), as well as in contentious proceedings before the Federal Patent Court (BPatG). Therefore, the patentability criterion of inventive step deserves the utmost attention.

An invention is considered new if it cannot be derived from any document or other publication prior to the filing or priority date. An invention involves an inventive step if it is not obvious to the person skilled in the art. In grant proceedings before the patent offices and in contentious proceedings before the Federal Patent Court, disagreements regarding the assessment of the novelty of a patent claim are rare. It can be clearly determined whether the subject matter of a patent claim is disclosed in a prior art reference or not.

However, the criterion of inventive step is difficult to assess in practice. This raises the question of what is considered obvious and what the skilled person would not have easily arrived at. With a skillful justification, it is sometimes still possible to obtain a patent grant in cases where a less suitable argumentation regarding inventive step would have had no chance. This book aims to provide practical tips to help overcome the hurdle of inventive step, or, in contentious proceedings before the Federal Patent Court, to successfully argue against the inventive step of a competitor's patent.

The term "inventive step" is commonly used to describe the patentability threshold under German and European patent law. Sometimes the term "inventive activity" is used. In US patent law, the term "non-obviousness" is customary. "Inventive activity" is understood in an economic and scientific context to describe creative work that leads to innovation. "Inventive activity" is also used as a collective term for research and development activities. For this reason, the term "inventive activity" can be regarded as a generic term for creative inventive activity, which may result in patents granted. This is the reason for chosing the term "inventive activity" for the title of the book. This technical book also imparts the theoretical foundations of patent law in order to provide a solid understanding for successfully establishing inventive step in accordance with the Patent Act.

Munich
in January 2025

Dr. Thomas Heinz Meitinger

Contents

About the Author

Dr. Thomas Heinz Meitinger is a German and European patent attorney. He is the managing director of Meitinger Patentanwalts GmbH. Meitinger Patentanwalts GmbH is a medium-sized patent law firm in Munich. After studying electrical engineering in Karlsruhe, he initially worked as a development engineer. Later positions included roles as production manager and technical director in medium-sized companies. Dr. Meitinger regularly publishes scientific articles and writes technical books on intellectual property law and technical topics. He gives lectures on patent, trademark, and design law. Dr. Meitinger holds the degrees Dipl.-Ing. (Univ.) and Dipl.-Wirtsch.-Ing. (FH). In addition, he holds the following master's degrees: LL.M., LL.M., MBA, MBA, M.A., and M.Sc.

List of Abbreviations

ABl. EPA	Official Journal of the European Patent Office
BeckOK	Beck's Online Commentaries
BeckRS	Beck case law (decision database from Beck-Online)
BGH	Federal Court of Justice
BGHZ	Decisions of the Federal Court of Justice in Civil Matters
BIPMZ	Journal for Patent, Design, and Trademark Matters
BPatG	Federal Patent Court
DPMA	German Patent and Trademark Office
EPA	European Patent Office
EPA G	Case Law of the Enlarged Board of Appeal of the European Patent Office
EPA T	Case Law of the Technical Boards of Appeal of the European Patent Office
EPG	Unified Patent Court (also UPC: Unified Patent Court)
EPÜ/EPC	European Patent Convention
GRUR	Industrial Property Protection and Copyright
GRUR-RS	Collection of Case Law
GRUR Int	GRUR International
Mitt.	Communications of the German Patent Attorneys
NJW	New Legal Weekly Journal
PatG	German Patent Act
RG	Imperial Court
RGZ	Collection of Decisions of the Imperial Court
Rn.	Margin number
VPP Rdbr	Newsletter of the Association of Industrial Property Protection Professionals (VPP)

List of Figures

The History of Inventive Step 1

Contents

To understand the criterion of inventive step, it is helpful to consider its historical development. The history of inventive step is characterized by a continuous effort to establish a patentability criterion that rewards only special creative achievements. Trivialities should not lead to patents.

1.1 The first German Patent Act of 1877

The first Patent Act of 1877 did not include inventive step as a requirement for patentability. The first Patent Act required only novelty and industrial applicability.[1] However, from the outset, the Patent Office and the Imperial Court accepted only those inventions as worthy of a patent that exceeded the average skill of a person skilled in the art.[2]

[1] Wikisource, https://de.wikisource.org/wiki/Patentgesetz, accessed on 13.10.2024.

[2] Benkard PatG/Asendorf/Schmidt/Tochtermann, 12th ed. 2023, PatG § 4 para. 7.

T. H. Meitinger, *Justification of the Inventive Activity*,
https://doi.org/10.1007/978-3-662-72735-5_1

1.2 Development up to the Patent Act of 1978

In the period from the first Patent Act of 1877 until the introduction of the Patent Act of 1978, there was no statutory regulation of the criterion of inventive step. During these decades, judicial development of the law and the granting practice of the Patent Office developed a set of tools for assessing the patentability of an invention, requiring novelty, inventive step, and technical progress.[3] A high level of inventive step or technical progress could compensate for a deficiency in the other. An invention with significant technical progress required only a low inventive step for grantability. Conversely, a high inventive step could make up for a low level of technical progress.[4] However, the criterion of technical progress lacked general acceptance, since it only becomes clear in retrospect in which direction a technology develops, and thus technical progress can only be confirmed in hindsight.[5]

1.3 Patent Act of 1978

The Patent Act of 1978 abolished technical progress as a criterion for patentability. In addition, the term "inventive step" replaced "level of invention." Technical progress could only serve as evidence of inventive step.[6]

[3] Ortwin Schulze, Technical Progress and Inventive Step, Mitteilungen der deutschen Patentanwälte, 1976, Issue 7/8, page 132.

[4] Beier, On the Historical Development of the Requirement of Inventive Step, GRUR 1985, 606, 615.

[5] Benkard EPÜ/Melullis/Koch, 4th edition 2023, EPÜ Article 52 para. 5.

[6] Benkard PatG/Asendorf/Schmidt/Tochtermann, 12th ed. 2023, PatG § 4 para. 15.

2 Fundamentals of Patent Law

Contents

Patent law is justified by natural law, according to which an invention is the property of the inventor. Ultimately, this would mean that an invention should permanently belong to the inventor and that the exclusive right should last forever. This would hinder the further development of technology.[1] The compromise is the temporal limitation of patent rights to a maximum of 20 years.[2]

The purpose of patent law is alternatively explained by the incentive theory, whereby a patent is intended to motivate the inventor to advance technology through intellectual creation and to invest time and money in research and development, since there is the

[1] Rogge/Melullis, Introduction, Benkard, Patent Act: PatG 12th edition 2023, para. 3.

[2] Section 16 Patent Act.

T. H. Meitinger, *Justification of the Inventive Activity*,
https://doi.org/10.1007/978-3-662-72735-5_2

prospect of legally protected profits.[3] The reward theory states that the applicant should be granted an exclusive right, so that the applicant has the opportunity to recoup their investments and, as a reward for their efforts, to earn a profit.[4] The restriction of free competition by the monopoly right of patents is thus justified.

Patents are granted exclusively for inventions with a technical character.[5] Inventions granted a patent must be published so that the public can study the patented technologies.[6] In this context, inventions in patents must be described in such a way that a skilled person can reproduce them.[7]

2.1 DPMA, EPO, BPatG and UPC

DPMA is the name for the German Patent and Trademark Office, headquartered in Munich. It is responsible for granting patents and registering utility models, trademarks, and designs for the territory of Germany.

EPO is the acronym for the European Patent Office, which was founded on October 7, 1977. It is responsible for granting European patents. The EPO is an intergovernmental organization headquartered in Munich. Other locations are in The Hague, Berlin, Vienna, and Brussels.

UPC is the acronym for the Unified Patent Court, which began operations on June 1, 2023. The Unified Patent Court is particularly responsible for nullity actions against European patents and unitary patents. It also has jurisdiction over infringement actions based on European patents and unitary patents.

A unitary patent (European patent with unitary effect or EU patent) provides uniform patent protection in the 18 EU member states that have ratified the Agreement on a Unified Patent Court. A joint unitary patent can be obtained for the following EU member states: Belgium, Bulgaria, Denmark, Germany, Estonia, Finland, France, Italy, Latvia, Lithuania, Luxembourg, Malta, Netherlands, Austria, Portugal, Romania, Slovenia, and Sweden. An important country in Europe where unitary patents have no effect is the United Kingdom.

[3] Rogge/Melullis, Introduction, Benkard, Patent Act: PatG 12th edition 2023, para. 3b.

[4] Rogge/Melullis, Introduction, Benkard, Patent Act: PatG 12th edition 2023, para. 1; BGH GRUR 1969, 534 (535)—Ski boot fastener.

[5] Section 1(1) Patent Act.

[6] BGH GRUR 1981, 734, 735—Erythronolid.

[7] Section 34(4) Patent Act; Article 83 EPC.

2.2 The Invention

An invention under the Patent Act is a technical teaching, similar to a recipe, for the manufacture of a device. The invention can be described in terms of process steps or features of the device. The invention must have a technical character.[8] An aesthetic design that serves no technical function is not eligible for patent protection.[9] The Design Act was created for the protection of designs.[10]

The central concept of patent law is the "invention." Only for an object or process to which the term "invention" applies is the Patent Act applicable. The Patent Act does not contain a legal definition of the term invention. Section 1(1) Patent Act at least makes it clear that an invention must have a technical character.[11]

The legislator has not defined when an invention exists. This "gap" was left by the legislator quite deliberately. The absence of a definition of invention is intended to allow this concept to be adapted to the respective state of the art.[12] The legislator has therefore assigned it to legal practice, in particular the patent office and the courts, to define for which products and processes the Patent Act is applicable.

A current example of the development of the meaning of the term invention or the required technicality is software patents.[13] With regard to the question of what is excluded as "software as such" under Section 1(3) No. 3 and (4) Patent Act and what is still patentable, there has been a consistent development of the law by the patent offices and courts.[14] In this respect, there has even been a temporarily divergent practice between the German and the European Patent Office.

---

[8] Section 1(1) Patent Act.

[9] Section 1(3) No. 2 Patent Act.

[10] German design law: Design Act; design law in the EU: previously: Community design, more recently: Union design.

[11] Beyer, Patent and Ethics in the Mirror of Technical Evolution, GRUR 1994, 541, para. 543.

[12] Bundesrat printed matter No. 14 1876/77 on Section 1; Becker, Maximilian: On the Freedom to Act Illegally, ZUM 2019, 636, 642.

[13] Lutz van Raden, the Informatics Dove—Considerations on the Patentability of Information-Related Inventions, GRUR 1995, 451.

[14] Andreas Wiebe, Patent Protection and Software Development—an Irreconcilable Contradiction? From the book "Open Source Yearbook 2004—Between Software Development and Social Model", 2004, pp. 277–292.

2.3 Technical Character

A patentable invention has a technical character.[15] Patent law is only accessible to inventions in the technical field. For example, designs are explicitly excluded from patent protection.[16]

2.4 Publications of the State of the Art

Inventive step is assessed against the background of the state of the art. For the assessment of inventive step, all technical teachings of the state of the art that were made available to the public before the filing or priority date of the relevant application may be taken into account.[17] The type or place of disclosure—oral, written, by presentation, domestic or foreign—does not matter. For utility models, the scope of the state of the art is slightly more limited compared to patent applications.[18]

The comparison with the state of the art serves to prevent monopolies for technical teachings that are already known or are obvious from the state of the art.[19] In particular, what is already known and what is obvious from it should be kept free from rights of prohibition.

"Making available to the public" can already occur by disclosure to a single expert who is not bound to secrecy.

2.5 Feasibility

An invention must be described in the patent application in such a way that it can be carried out by a person skilled in the art. To demonstrate feasibility, the description should include at least one example for each particular embodiment of the invention. In particular, the European Patent Office is increasingly requiring that corresponding examples be described in the patent application to provide clear evidence that the specific embodiments of the invention have actually been created, and not merely to "preemptively" claim an entire technical field.[20]

[15] Section 1(1) Patent Act.

[16] Section 1(3) No. 2 Patent Act.

[17] Section 3(1) sentence 2 Patent Act.

[18] Section 3 sentence 2 Utility Model Act. For utility models, oral disclosures and uses abroad do not constitute relevant prior art.

[19] Section 4 sentence 1 Patent Act.

[20] BeckOK PatR/Einsele, 33rd ed. 15.7.2024, EPC Article 56 para. 1c.

The patent office particularly suspects a lack of feasibility when features are linked with "and/or". In such cases, the description should include embodiments that exhibit the first, the second, and both features that are connected by "and/or".

2.6 Novelty

The requirements for granting a patent are, in particular, novelty, inventive step, and industrial applicability of the invention.[21] The novelty examination is intended to ensure that only those inventions are patented which enrich technical knowledge with a previously unknown embodiment. In the novelty examination, an individual comparison is made, in which the invention is compared separately with each device or process of the prior art. There is no overall consideration of the prior art.[22]

2.7 Inventive Step

The term "inventive step" is an indeterminate legal concept. Its determination requires an evaluative assessment.[23] The assessment depends on the prior art found, the person skilled in the art, and the interpretation of the claims.

The difference between the novelty examination and the assessment of inventive step lies in the fact that, in the novelty examination, only a single document or a single disclosure of the prior art is compared with the invention. In the novelty examination, there is no mosaic-like overall consideration of several documents. In contrast, the assessment of inventive step involves an overall consideration of several documents of the prior art.[24] However, the prior art is not to be combined arbitrarily. Rather, the question is which documents the skilled person would have to combine and whether there was any motivation to do so. If suitable documents existed in the prior art and the skilled person had reason to combine them, it must then be considered what arguments support or oppose combining the technical teachings of these documents.[25]

It is only ever possible to positively prove the absence of an inventive step, since a document may always be found later that calls the inventive step into question. The finding that

[21] Section 1 paragraph 1 Patent Act.

[22] Benkard PatG/Asendorf/Schmidt/Tochtermann, 12th edition 2023, PatG § 4 para. 20.

[23] Federal Court of Justice decision of 17.1.1995, X ZB 15/93, GRUR 1995, 330—Electrical Plug Connection.

[24] BeckOK PatR/Einsele, 33rd ed. 15.7.2024, EPC Art. 56 para. 1a; BGH GRUR 54, 24—Multiple Clamp; BGH GRUR 64, 167—Writing Instrument; BGH GRUR 74, 208—Power Supply Device.

[25] BGH GRUR 2013, 164—Guide Rail; Meier-Beck GRUR 2013, 1177.

an invention is based on an inventive step is therefore a presumption.[26] The same applies to the novelty of an invention. In contrast, a lack of inventive step and lack of novelty can be definitively established. A subsequently discovered document of the prior art can no longer change this.

2.8 Industrial Applicability

A prerequisite for patenting is that the invention is industrially applicable.[27] However, this criterion is almost always met, as it is hardly conceivable that there would be no possibility of commercial exploitation for a product. This criterion therefore plays no role in the granting of patents.

2.9 "Inventive Activity" in Patents versus "Inventive Step" in Utility Models

The required inventive level in patent law is referred to as "inventive activity"[28]. The required inventive level in utility model law is called "inventive step"[29]. The legislator intended, by using different terms for the respective required inventive level, to make clear that these are different criteria with different requirements. The legislator wanted to provide, through the utility model, a form of protection for "small" inventions. In particular, a lower creative achievement should suffice for the legal validity of a utility model compared to a patent.[30]

Until 2004, the 5th Board of Appeal of the Federal Patent Court, which is responsible for utility models, decided in accordance with the legislator's intention and accepted the legal validity of utility models whose invention would not have met the standard of inventive activity under patent law. According to Section 4 sentence 1 Patent Act, an inventive activity is present if the invention is not obvious in light of the prior art. The invention of a utility model could therefore even have been obvious in view of the prior art. However, according to the prevailing view at the time, the invention of a utility model should at least not result directly from the specialist knowledge of the skilled person and any routine experiments.[31]

[26] Benkard PatG/Asendorf/Schmidt/Tochtermann, 12th edition 2023, PatG § 4 para. 38.

[27] Section 1 paragraph 1 Patent Act.

[28] Section 4 sentence 1 Patent Act.

[29] Section 1 paragraph 1 Utility Model Act.

[30] Federal Patent Court GRUR 2006, 489—Impact Tool.

[31] Federal Patent Court GRUR 2004, 852—Material Strip Packaging.

The Utility Model Board of the Federal Patent Court was aware of the vague distinction between the inventive step in utility model law and the inventive activity in patent law. The Board of Appeal therefore allowed an appeal on points of law in a relevant case, and the Federal Court of Justice had to decide on the definition of the inventive step.[32]

The Federal Court of Justice first established that the inventive level of a protective right is a qualitative question and that, therefore, no quantitative deductions can be made from the criterion of inventive activity to arrive at the minimum standard of an inventive step.[33]

The Federal Court of Justice also established that something obvious cannot form the basis for a monopoly right. It is not acceptable for trivial innovations, which arise for a skilled person from their expertise, to already lead to an exclusive right. On the other hand, from the perspective of the Federal Court of Justice, the definition of inventive activity already represents a lower limit, since something obvious cannot be inventive. It follows that the inventive step must have at least the same inventive level, thereby making the inventive step in utility model law de facto equivalent to the inventive activity in patent law.[34]

The Federal Court of Justice saw a need to prevent trivial protective rights, which could be used to monopolize the area around the prior art that a skilled person can use through their expertise and knowledge without being inventive. In this way, from the perspective of the Federal Court of Justice, normal technological development continues to be enabled.[35]

In particular, the Federal Court of Justice argued in its "Demonstration Cabinet" decision that the same requirements must be applied to an inventive step[36] and to inventive activity[37], since both types of protective rights confer the same rights.[38] However, the "Demonstration Cabinet" decision is considered controversial.[39]

---

32 Winterfeldt, Case Law Overview 2005, GRUR 2006, 441, 459; Winterfeldt/Engels, Case Law Overview 2006, GRUR 2007, 537, 546.

33 Benkard PatG/Engel 12th edition 2023, GebrMG § 1 para. 16.

34 Meier-Beck Case Law Overview for 2006, GRUR 2007, 913, 914–915.

35 Benkard PatG/Engel 12th edition 2023, GebrMG § 1 para. 16c-d.

36 Section 1 paragraph 1 Utility Model Act.

37 Section 4 sentence 1 Patent Act.

38 Federal Court of Justice GRUR 2006, 842 = Mitteilungen der deutschen Patentanwälte, 2006, 512—Demonstration Cabinet.

39 Mes, 5th edition 20.020, PatG § 4 para. 88.

2.10 No Patent Protection

Patents are not granted for discoveries, scientific theories, mathematical methods, aesthetic creations, schemes, rules and methods for mental activities, for games or for business activities, as well as software.[40] However, software is only not eligible for protection insofar as it does not have a technical character.[41]

2.11 No Right to Prohibit

Patents cannot be used to prevent private use. Non-commercial application of a technical teaching is permitted.[42] It is also allowed to conduct experiments to learn about the invention.[43] Methods for surgical or therapeutic treatment of the human or animal body and diagnostic methods performed on the human or animal body are not patentable. Corresponding patents are unenforceable. Such methods may be used.[44] There is also no patent infringement if a pharmacist prepares a medicinal product as an individual preparation.[45]

[40] Section 1 paragraph 3 Patent Act.
[41] Section 1 paragraph 4 Patent Act.
[42] Section 11 no. 1 Patent Act.
[43] Section 11 no. 2 Patent Act.
[44] Section 2a paragraph 1 no. 2 Patent Act.
[45] Section 11 no. 3 Patent Act.

The Person Skilled in the Art in Patent Law

3

Contents

The patent law expert is an average skilled person with average technical knowledge and ability. The patent law expert serves as the standard for deciding whether an invention is based on an inventive step.[1] Everything that the skilled person knows due to their technical knowledge or can deduce using their technical skills is considered obvious and therefore not inventive.[2]

[1] Benkard PatG/Asendorf/Schmidt/Tochtermann, 12th ed. 2023, PatG § 4 para. 62; Osterrieth GRUR 2021, 310, 311.

[2] § 4 sentence 1 Patent Act; Article 56 sentence 1 EPC.

T. H. Meitinger, *Justification of the Inventive Activity*,
https://doi.org/10.1007/978-3-662-72735-5_3

The patent law expert is a fictional construct that must always be used when a specific technical matter is to be assessed.[3] The skilled person must be taken into account when considering the legal validity of a patent claim and when determining whether a specific infringement of a claim has occurred. The skilled person is to be considered when assessing prior art documents, when evaluating the reproducibility of an invention (§ 34 (4) PatG), when assessing novelty (§ 3 PatG), when evaluating inventive step (§ 4 PatG), when interpreting patent claims to determine the scope of protection (§ 14 PatG), and when assessing an equivalent form of infringement.[4]

The concept of the skilled person is a central term in patent law. The "skilled person" is the decisive standard of assessment.[5] However, determining the skilled person presents a significant challenge.[6] In principle, the determination of the skilled person depends on the technical field to which the invention belongs.[7] The patent law expert is not a real person, but a fictional character.[8] The patent law expert is considered the average expert in their technical field, who has received the usual training and has acquired practical experience through their professional activity.[9]

The skilled person in patent law is crucial for the correct technical understanding of inventions, for assessing novelty and inventive step, and especially for determining the scope of protection of a patent when evaluating a possible patent infringement.[10]

In particular, the skilled person is the pivotal point in assessing the patentability of an invention. An assessment must always be made from the perspective of the skilled person. The Patent Act itself mentions the skilled person in § 34 paragraph 4 Patent Act, which determines the executability of a patent.[11]

A patent or a patent application contains a technical teaching that is addressed to a technical expert with average knowledge and ability, not to a layperson. Furthermore, the

---

[3] BGH GRUR 2004, 1023, 1025—device for separating items from the base; BGH GRUR 2006, 663—prepaid telephone calls.

[4] Dreiss, The average skilled person as a standard for sufficient disclosure, patentability, and patent interpretation, GRUR 1994, 781, 782.

[5] Osterrieth, The skilled person in patent law, GRUR 2021, 310.

[6] Klett, The reasonably attentive consumer and the reasonably well-trained skilled person, GRUR 2001, 549, 554; Benkard PatG/Asendorf/Schmidt/Tochtermann, 12th ed. 2023, PatG § 4 para. 64.

[7] Benkard PatG/Asendorf/Schmidt/Tochtermann, 12th ed. 2023, PatG § 4 para. 64; BGH GRUR 1959, 532, 536 f.—Electromagnetic stirring device; BGH Liedl 1974/1977, 69, 78—Target for shooting; BGH BeckRS 2009, 23384 para. 31 = BeckRS 2009, 23384.

[8] BGH GRUR 2004, 1023—device for separating items from the base; GRUR 2006, 663 = BlPMZ 2006, 320—prepaid telephone calls.

[9] BeckOK PatR/Fitzner/Metzger, 29th Edition 15.4.2023, PatG § 3 para. 8.

[10] Osterrieth, The skilled person in patent law GRUR 2021, 310.

[11] § 34 paragraph 4 Patent Act: "The invention must be disclosed in the application in such a clear and complete manner that a skilled person can carry it out."

skilled person is mentioned in § 4 sentence 1 Patent Act.[12] Although the skilled person is generally a fictional character, the patent law expert can also exist as a specific natural person.[13]

The patent law expert is familiar with the entire state of the art in their technical field. In addition, they possess basic knowledge of adjacent technical fields. However, the patent law expert is not creative; they can only derive what is obvious from their technical knowledge.[14]

The scope of protection of a patent is determined by § 14 Patent Act. The skilled person is not mentioned in this section. However, the skilled person is referenced in the Protocol on the Interpretation of Article 69 EPC,[15] which, according to Article 164 paragraph 1 EPC, forms part of the European Patent Convention (EPC). The perspective of the skilled person is therefore also decisive when assessing a patent infringement.

The notional person skilled in the art under patent law is an average skilled person,[16] who works in the technical field of the invention[17] and possesses average knowledge and skill[18]. The skilled person is defined by the technical problem at hand and not with regard to the invention itself or the inventive step required to achieve the invention.[19] The state of the art cannot be used to define the skilled person.[20] The skilled person can be understood as being part of a team, whereby he does not possess the specialized knowledge

[12] § 4 sentence 1 Patent Act: "An invention is considered to be based on an inventive step if it is not obvious to the skilled person from the state of the art."

[13] BGH GRUR 2004, 1023—device for separating items from the base; Kulhavy, Mitteilungen der deutschen Patentanwälte, 2011, 179; Klett GRUR 2001, 549, 552; BGH GRUR 2018, 390—thermal energy management; GRUR 2010, 602—joint arrangement; Meier-Beck, Mitteilungen der deutschen Patentanwälte (Mitt.) 2005, 529; Niedlich FS König, 2003, 399; Osterrieth, The skilled person in patent law, GRUR 2021, 310, 312–313.

[14] BeckOK PatR/Einsele, 33rd Ed. 15.7.2024, EPC Article 56 para. 6.

[15] EPO, https://www.epo.org/law-practice/legal-texts/html/epc/2020/d/ma2a.html, accessed on 25.10.2024.

[16] BGH GRUR 1995, 330—Electrical Plug Connection; GRUR 2004, 1023—Device for Separating Items from the Bottom; GRUR 2006, 663—Prepaid Telephone Calls; BeckRS 2005, 12061—Folding Door Unit; GRUR 2018, 390—Thermal Energy Management.

[17] BGH GRUR 1978, 37—Stock Exchange Clip; GRUR 1962, 290—Carrier Pigeon Travel Cabin II; GRUR 2005, 1023—Shopping Cart II; GRUR 2011, 1109—Tire Sealant; GRUR 2010, 513—Sectional Overhead Door II; BPatG Mitt. 1984, 213.

[18] BGH BlPMZ, 1991, 159—Adhesive Bandage; GRUR 2009, 929—Abrasive Grain; GRUR 2004, 272—Diabetes Ratio.

[19] BGH GRUR 2018, 390—Thermal Energy Management; GRUR 2010, 602—Joint Arrangement; GRUR 2016, 921—Pemetrexed.

[20] BGH GRUR, 2009, 382—Olanzapine; GRUR 2009, 1039—Fish Bite Indicator; GRUR 2018, 390—Thermal Energy Management; GRUR 2017, 498—Knitted Shoe Upper.

of other experts, but can consult them if necessary, without this constituting an inventive step.[21]

The skilled person possesses technical knowledge and technical skill. Technical skill, i.e., the ability of the skilled person to understand a technical subject matter through their own considerations and suitable experiments, is difficult to determine. In fact, a change in the qualification of the skilled person can be observed over the years, with the notional person skilled in the art under patent law having lost some creative qualification.[22]

3.1 Role of the Skilled Person

It must be ensured that patent claims are always interpreted consistently. For this purpose, a binding instance for the interpretation and understanding of patent claims must be established. This function is fulfilled by the definition of the notional person skilled in the art under patent law, who is an objectively judging person for the assessment of a patent and the relevant state of the art.[23] This ensures an objective interpretation of patent claims.[24] The creation of the notional average skilled person under patent law thus provides a reliable basis for assessing inventive step and thereby promotes legal certainty.[25] The influence of individual knowledge and abilities can be excluded by defining the notional average skilled person. The skilled person possesses the general technical knowledge and skill customary in the relevant technical field.[26]

The notional person skilled in the art under patent law determines how claims are to be understood. A claim formulation may always include embodiments that are technically nonsensical. The skilled person is needed to recognize such technically nonsensical embodiments and to exclude them from the scope of protection of the relevant claim. By defining the notional person skilled in the art, it is prevented that a claim is understood

[21] BGH GRUR 2007, 404 or BGHZ 170, 215—Carvedilol II; BGH GRUR 2010, 123—Escitalopram; EPO T 460/87—Soluble Polyglycols/Viscosud; EPat 99/89—Circuit Arrangement for Optical Display of State Variables/Robert Boschmann GmbH.

[22] Dreiss, The Average Skilled Person as a Standard for Sufficient Disclosure, Patentability, and Patent Interpretation, GRUR 1994, 781, 783.

[23] BGH GRUR 2018, 390 para. 31—Thermal Energy Management.

[24] Timmann, § 3. Interpretation and Scope of Protection of Patents in Haedicke/Timmann, Handbook of Patent Law, 2nd edition 2020, para. 26.

[25] Timmann, § 3. Interpretation and Scope of Protection of Patents in Haedicke/Timmann, Handbook of Patent Law, 2nd edition 2020, paras. 8–12; Benkard PatG/Asendorf/Schmidt/Tochtermann, 12th ed. 2023, PatG § 4 para. 62.

[26] BGH judgment of 7 September 2004, X ZR 255/01, GRUR 2004, 1023, 1025—Device for Separating Items from the Bottom.

differently depending on the level of education or knowledge, which would otherwise lead to legal uncertainty.[27]

The tasks of the notional person skilled in the art in assessing inventive step are as follows:

- Determining the subject matter claimed by the patent. In this context, the skilled person interprets the patent claims in light of their technical knowledge and skill.
- Determining the relevant state of the art. The question arises as to which documents the skilled person would have consulted to solve the technical problem.
- Determining the disclosure content of the state of the art. How would the skilled person have understood the state of the art, and what technical teachings would they have derived from it?
- Obviousness. Would the skilled person have combined the relevant documents of the state of the art to arrive at the invention, or were the respective documents technically incompatible, or did the respective technical teachings perhaps even contradict each other? Would additional skills or knowledge have been required for the skilled person to arrive at the invention?

3.2 Significance of the Skilled Person

The skilled person is crucial for the interpretation of claims. The determination of the scope of protection of the claims forms the basis of a patent infringement proceeding. Only with the technical knowledge and skill of the skilled person can a correct understanding of the claims be achieved as the starting point for any decision under patent law.[28]

When interpreting claims, one should not adhere to a linguistic or logical content, but rather to the technical understanding that is immediately and unambiguously apparent to the skilled person.[29] The skilled person is the decisive instance for assessing whether a patent is legally valid and thus whether its enforcement is justified.[30]

The skilled person is always defined in German grant proceedings before the Patent Office and in appeal and nullity proceedings before the Federal Patent Court. In contrast, the skilled person is, surprisingly, not explicitly defined in proceedings before the European Patent Office.

---

[27] Benkard EPC/Scharen, 4th ed. 2023, EPC Art. 69 para. 11.

[28] BGH GRUR 2018, 390 paras. 31 and 37—Thermal Energy Management; Nägerl, § 4. Distinction from the State of the Art—Novelty and Inventive Step, Haedicke/Timmann, Handbook of Patent Law 2nd edition 2020 para. 46.

[29] Nägerl, § 4. Distinction from the State of the Art—Novelty and Inventive Step, Haedicke/Timmann, Handbook of Patent Law 2nd edition 2020 para. 49.

[30] Loth, PatG § 14 [Scope of Protection] in BeckOK Patent Law, Fitzner/Kubis/Bodewig, 30th edition, as of 15.10.2023, para. 56.

3.3 Determination of the Skilled Person

The skilled person is determined based on the technical field of the invention under examination. Care must be taken not to use the specific technical problem or parts of the invention to define the skilled person. If, in a retrospective view, the skilled person is placed too close to the invention, the invention, although patentable, suddenly appears obvious.[31] Nor may the skilled person be defined based on the closest prior art.[32]

3.4 Fictitious Skilled Person

The skilled person is, as a rule, a fictitious person. When determining the skilled person, it is therefore not necessary to refer to specific, real-life experts. This ensures legal certainty, as the relevant professional thinking of the appropriate average skilled person can be determined without being restricted by the requirement to present a real-life expert.[33] The patent law skilled person is a legal construct created to objectively understand and assess an invention.[34] The subjective achievement of the inventor or what they "actually" intended to convey is irrelevant.

The patent law skilled person is not a real but a fictitious person. In this respect, a certain similarity can be seen to artificial intelligence, which also operates in the virtual realm. However, it can by no means be concluded from this that artificial intelligence could replicate the current patent law skilled person.[35] The skilled person cannot be replaced by artificial intelligence, as artificial intelligence cannot replicate the "motivation" of the skilled person to combine or not to combine documents from the prior art. Rather, for artificial intelligence, it is irrelevant whether documents are combinable or not, since, due to its high computing capacity, it can easily combine all relevant documents. If artificial intelligence were to be defined as the patent law skilled person, the could-would test could be dispensed with.[36]

[31] Dreiss, The Average Skilled Person as a Standard for Sufficient Disclosure, Patentability, and Patent Interpretation, GRUR 1994, 781, 786.

[32] BGH GRUR 2009, 382—Olanzapine; GRUR 2009, 1039—Fish Bite Indicator; GRUR 2018, 390—Thermal Energy Management; GRUR 2017, 498—Knitted Shoe Upper.

[33] BeckOK PatR/Loth, 29th ed. 15.7.2023, PatG § 14 para. 58; BGH GRUR 2004, 1023, 1025—Device for Separating from the Bottom.

[34] Osterrieth, Part 4. Subject Matter, Requirements, and Effect of Patent Protection, Osterrieth, Patent Law, 6th edition 2021, para. 470.

[35] Loth, PatG § 14 [Scope of Protection] in BeckOK Patent Law, Fitzner/Kubis/Bodewig, 30th edition, as of 15.10.2023, para. 56b.

[36] Benkard PatG/Asendorf/Schmidt/Tochtermann, 12th ed. 2023, PatG § 4 paras. 34–37.

The fact that the patent law skilled person is not a real-life expert but a legal construct can also be seen from the fact that the patent law skilled person is not assigned the expertise and skills of a real-life expert, but rather average expertise and skills.[37]

3.5 Average Skilled Person

The patent law skilled person is assumed to be an average skilled person. The average skilled person possesses average expertise and average ability in their technical field and is considered a cross-section of all professionals active in the technical field on the priority date who would have been entrusted with solving the technical problem.[38]

3.6 Technical Knowledge

The general state of knowledge among experts in a technical field is referred to as technical knowledge. The technical knowledge of the patent law skilled person includes general knowledge and familiarity with the relevant state of the art.[39] The skilled person is assumed to possess technical knowledge that encompasses disclosures from the state of the art, field-specific knowledge, and general technical knowledge.[40]

Technical knowledge includes the technical fundamentals acquired by the skilled person through education, as well as specialized knowledge gained through professional activity in the technical field of the invention. The skilled person therefore possesses both knowledge learned through education and experiential knowledge. Technical knowledge does not have to be identical to the state of the art. The state of the art encompasses more than what is attributable to technical knowledge.[41] Conversely, technical

[37] Benkard PatG/Asendorf/Schmidt/Tochtermann, 12th ed. 2023, PatG § 4 para. 18.

[38] BGH GRUR 2004, 1023, 1025—Device for Separating from the Bottom; BGH GRUR 2018, 390 para. 31—Thermal Energy Management.

[39] Keukenschrijver in: Busse/Keukenschrijver, Patent Act, 9th ed. 2020, § 4 (Inventive Step), para. 93; BGH GRUR 1978, 37 f. Börsenbügel; BGH BlPMZ 1989, 133 Gurtumlenkung; BGH 18.8.1983 X ZR 17/82; BPatGE 32, 109 = BlPMZ 1991, 349, 351; EPO T 195/84 OJ EPO 1986, 121 = GRUR Int 1986, 545 f.; EPO T 426/88 OJ EPO 1992, 427, 433 = GRUR Int 1993, 161 Combustion Engine; EPO T 206/83 OJ EPO 1987, 5 = GRUR Int 1987, 170 Herbicides; EPO T 99/85 OJ EPO 1987, 413 = GRUR Int 1988, 251 Diagnostic Agent; EPO T 580/88; EPO T 766/91; EPO T 378/93 and EPO T 590/94; EPO T 890/02 OJ EPO 2005, 496 = GRUR Int 2005, 1030 Chimeric Gene; Benkard para. 75; Szabo Mitt 1994, 225, 232 f.; Kraßer/Ann § 18 para 47 fn 75; RG GRUR 1938, 709, 713 Silencer; BGH Liedl 1967/68, 204, 218 Core Brake; BGHZ 133, 79 = GRUR 1996, 862, 865 Arc Segment; BGHZ 160, 204 = GRUR 2004, 1023 Bottom-Side Singulation Device; BPatG 22.11.2002 17 W (pat) 59/00; BPatG 20.1.2015 3 Ni 18/13 (EP).

[40] Benkard PatG/Asendorf/Schmidt/Tochtermann, 12th ed. 2023, PatG § 4 para. 69.

[41] Gramm: The State of the Art and Technical Knowledge, GRUR 1998, 240, 241.

knowledge includes technical teachings that are not part of the state of the art.[42] For example, the experiential knowledge of a skilled person includes the understanding that a plug connection requires a plug as a complementary counterpart.[43]

The content of a document from the state of the art includes everything that the skilled person would "read between the lines." In particular, the skilled person would "read between the lines" the use of so-called "notorious" (interchangeable) means. Notorious means include, for example, coils for generating a magnetic field to deflect electrons. Therefore, when a description refers to the deflection of electrons with a magnetic field, the skilled person will also think of coils. Technical skill can be defined as everything, no matter how laborious, that does not yet constitute an inventive step under § 4 sentence 1 of the Patent Act.[44] Findings resulting from standard experimental procedures are also attributable to the technical knowledge of the skilled person.[45]

The technical knowledge to be considered includes not only the present knowledge of the skilled person, but also the technical knowledge available worldwide in the technical field of the invention.[46]

Technical knowledge is derived from technical literature, such as textbooks, reference works, and technical journals.[47] Technical knowledge also results from practical professional activity. In addition, the skilled person possesses general technical knowledge.[48] It is assumed that the skilled person is at least generally familiar with the overarching technical field and with technical fields adjacent to their own.[49] Detailed knowledge of adjacent technical fields is not expected of a skilled person.[50]

However, the technical knowledge of the skilled person is not limited to their own technical field. Through their education, the skilled person has acquired general technical knowledge. In particular, the skilled person is familiar with related technical fields that solve similar technical problems.[51]

---

[42] Gramm: The State of the Art and Technical Knowledge, GRUR 1998, 240, 242.

[43] Ann, § 17. Novelty Ann, Patent Law, 8th edition 2022, para. 29.

[44] Dreiss, The Person Skilled in the Art as a Standard for Sufficient Disclosure, Patentability, and Patent Interpretation, GRUR 1994, 781, 784–785.

[45] Benkard PatG/Asendorf/Schmidt/Tochtermann, 12th ed. 2023, PatG § 4 para. 72; BGH GRUR-RS 2019, 39.560 para. 121; BGH BeckRS 2016, 15.771 para. 40; GRUR 2006, 666 para. 56—Stretch Film Hood; BGH BlPMZ 1966, 234 (235)—Scanning Method.

[46] BeckOK PatR/Einsele, 29th ed. 15.7.2023, PatG § 4 para. 36.

[47] BGH GRUR 1998, 895 (897)—Rainwater Basin; BPatGE 34, 264.

[48] BGH BlPMZ 1989, 133—Belt Deflection under II, 3.

[49] BGH GRUR 1997, 272 (273)—Swing Lever Lock; BGH BeckRS 2008, 2725 para. 23.

[50] EPO OJ 1997, 24; Benkard PatG/Asendorf/Schmidt/Tochtermann, 12th ed. 2023, PatG § 4 para. 71.

[51] BGB BeckRS 2000, 4931 para. 25; BGH GRUR 1963, 568, 569; BGH GRUR 1969, 182, 183; BGH GRUR 1986, 372, 374; BGH GRUR 1986, 798, 799.

3.7 Professional Competence

The skilled person possesses professional competence, which the skilled person has acquired through their specific work routine and self-conducted experiments. This professional competence represents the skilled person's ability to deal creatively with a new situation.[52]

In addition to technical knowledge, the skilled person therefore possesses creative ability, referred to as professional competence.[53] However, the fictional patent law skilled person is only attributed with a creative professional competence that does not suffice for inventive activity required to obtain a patent.[54] The patent law skilled person is therefore only capable of purely routine and constructive measures.[55]

The fictional patent law skilled person is assumed to be someone to whom one would entrust development activities in the relevant technical field. The patent law skilled person is expected to improve or simplify known devices and methods.[56] In doing so, they apply the methods customary in their technical field to simplify and make existing solutions more cost-effective.[57]

3.8 Determination of the Skilled Person by the Judge Involved

The ultimately binding determination of the applicable skilled person is made by the judge in opposition appeal, nullity proceedings, or infringement proceedings. A judge is not a subsumption machine, but rather incorporates individual assessments and experience.[58]

---

[52] BeckOK PatR/Einsele, 29th ed. 15.7.2023, PatG § 4 paras. 32–35; BGH GRUR 1986, 372—Platelet Counting; BlPMZ 1966, 234—Scanning Method; GRUR 2006, 666—Stretch Film Hood; BlPMZ 2009, 432—Airbag Trigger Control; GRUR 2012, 378—Installation Device II.

[53] Kraßer/Ann § 18 para. 46; Benkard PatG/Asendorf/Schmidt/Tochtermann, 12th ed. 2023, PatG § 4 para. 34; Benkard-EPÜ3 Art 56 para. 53; BGH 23.1.2007 X ZB 3/06.

[54] EPO T 39/93 OJ EPO 1997, 134, 149 f. = GRUR Int 1997, 741 Polymer Powder.

[55] BGH GRUR 1954, 107, 110 Multiple Clamp; BGH GRUR 1954, 258 f. Coal Mining; BGH GRUR 1956, 73, 76 California Shoes; BGH GRUR 1961, 529, 533 Radiation Apparatus; BGH GRUR 1962, 350 Triple Mirror Reflector; BGH Liedl 1967/68, 171, 192 Self-Closing Valve; BGH Liedl 1971/73, 261, 274 Cough Cap; BGH Liedl 1974/77, 50, 62 Collapsible Table; BGH Bausch BGH 1999–2001, 157 Circulating Pump for Household Appliances; BGH 17.11.2009 X ZR 49/08; EPO T 170/87 OJ EPO 1989, 441, 444 = GRUR Int 1990, 223 Flue Gas Cooler; see also the case law of the RG 8th ed.; Schulte para. 109; Benkard para. 112; Benkard-EPÜ1 Art 56 paras. 55, 131; cf. BPatG GRUR 1989, 745 f.; BPatG 8.3.2001 11 W (pat) 64/00; Austrian Patent Office öPBl 2001, 104, 106, Utility Model Matter.

[56] Benkard PatG/Asendorf/Schmidt/Tochtermann, 12th ed. 2023, PatG § 4 para. 72.

[57] BGH Bausch 1994/1998, 291 (296)—Collection Station; BGH Bausch 1994/1998, 168 (174)—Single-Phase Synchronous Motor.

[58] Winkelmann: Decision-Making by Artificial Intelligence in the Judiciary, LTZ 2022, 163, 165.

3.9 Team of Skilled Persons

The skilled person may consult a second skilled person if the technical problem concerns a second technical field.[59] The necessity of involving a second skilled person does not automatically render an invention inventive. In this case, the combined technical knowledge and professional competence of both experts constitute the knowledge and competence of the fictional patent law skilled person.[60] However, if it was not to be expected that a solution to the technical problem could be found in the other technical field, the involvement of a second skilled person is not permissible.

If there are no technological similarities between two technical fields, they are not considered neighboring technical fields. The skilled person will not take into account knowledge from a remote field.

3.10 Technical Neighboring Fields

The patent law skilled person in a technical field has fundamental knowledge of technical neighboring fields, but does not know all relevant documents from the neighboring field. If the skilled person can recognize that a solution to their problem may be found in a neighboring field, the essential knowledge of the neighboring field is attributed to the skilled person.[61]

3.11 Further Development of the Skilled Person

The concept of the skilled person, like the concept of invention, must be adapted to technological development.[62] With the emergence of artificial intelligence and its application by inventors in certain technical fields, the use of AI tools must also be permitted for the patent law skilled person.[63]

[59] BGH GRUR 78, 37 Stock Exchange Clip; BGH GRUR 83, 64 Reclining Furniture; BGH GRUR 10, 41 Diode Lighting.

[60] BGH GRUR 86, 798 Discharge Device for Bulk Material.

[61] BGH GRUR 89, 133 Belt Deflection.

[62] Meitinger, The Transhumanist Skilled Person, Mitteilungen der deutschen Patentanwälte, 115, March 2024, 109.

[63] Nägerl/Neuburger/Steinbach: Artificial Intelligence: Paradigm Shift in the Patent System, GRUR 2019, 336, 338.

Interpretation of the Claims

4

Contents

The patent holder may, by virtue of the right to prohibit under § 9 sentence 2 of the Patent Act, exclude any unauthorized third party from any form of use of the patented invention.[1] The items described by the claims define the scope of protection and establish a right to prohibit. The scope of protection of the claims is to be determined by interpretation. Therefore, an interpretation of the claims is mandatory prior to any patent enforcement.[2]

Patent enforcement is based on a claim for injunctive relief. An enforceable claim for injunctive relief by a patent holder or a third party, such as an exclusive licensee, exists if

[1] BGH GRUR 1973, 208, 209—Neues aus der Medizin; BGH GRUR 1980, 241—Rechtsschutzbedürfnis; BGH GRUR 1992, 318, 319—Jubiläumsverkauf; Mes, 5th ed. 2020, PatG § 139 para. 58.

[2] Timmann, § 3 Interpretation and Scope of Protection of Patents, Haedicke/Timmann, Handbook of Patent Law, 2nd edition 2020, para. 5.

T. H. Meitinger, *Justification of the Inventive Activity*,
https://doi.org/10.1007/978-3-662-72735-5_4

a patent claim has been infringed or if there is a serious likelihood that an infringement is imminent.

The patent claim is the decisive basis for determining the scope of protection of a patent.[3] The description and the drawings must always be used to support the interpretation of the patent claims, and not only in cases of ambiguity in the claims. The interpretation of a claim is independent of whether a patent infringement or the validity of a patent is being assessed.

In determining the scope of protection, only the independent patent claims are to be considered. The dependent claims, which refer to an independent claim, merely describe particular embodiments whose scope of protection is already encompassed by the scope of the independent claims.[4]

The meaning of the claims is to be determined by interpretation. Interpretation is carried out from the perspective of the skilled person.[5] Technical understanding is required to determine the patented technical teaching. According to § 14 sentence 2 of the Patent Act[6] the claims are to be interpreted in light of the description and the drawings.

The patent claims are to be interpreted by the skilled person in such a way that illogical and technically nonsensical embodiments are disregarded. The skilled person should, through structuring and a holistic view of the features, arrive at technically meaningful embodiments.[7] The claims are therefore not to be deliberately misunderstood by the skilled person, but rather interpreted with their technical knowledge and expertise so that the wording of the claim yields not nonsensical, but meaningful technical solutions.[8] It is important that the meaning of the words of the claim is not simply added together, but that the substantive content is developed based on the knowledge and expertise of the skilled person.

Determining the scope of protection of the claims is central to clarifying a patent infringement, and thus to patent enforcement.[9] The determination of the scope of protection of a claim must follow the principles of legal certainty and the requirement of appropriate reward. Legal certainty exists if any third party can clearly determine from the wording of the claim which embodiments are protected. Furthermore, a patent is

---

[3] Article 69 EPC in conjunction with the Protocol on the Interpretation of Article 69 EPC.

[4] Nägerl, § 4. Distinction from the State of the Art—Novelty and Inventive Step, Haedicke/Timmann, Handbook of Patent Law, 2nd edition 2020, para. 45.

[5] BGH GRUR 2016, 1031—Wärmetauscher: strict requirements for granting a sell-off period in the event of patent infringement.

[6] Analogous: Article 154 EPC.

[7] BGH 29.6.2010—X ZR 193/03, GRUR 2010, 858 para. 13—Crimpwerkzeug III; BGH 22.12.2009—X ZR 56/08, GRUR 2010, 214 para. 29—Kettenradanordnung II; Meier-Beck GRUR 2011, 857, 862.

[8] Benkard EPC/Wieser/Kinkeldey, 4th ed. 2023, EPC Art. 84 para. 12.

[9] Osterrieth, Part 6. Patent Infringement, Osterrieth, Patent Law, 6th edition 2021, para. 855.

intended to serve as a reward for the creation of the invention and the publication of the technical teaching. A revolutionary invention may be granted a broader scope of protection compared to a so-called "trivial patent".[10] In this way, the case law can achieve an appropriate scope of protection.[11]

However, it is the responsibility of every applicant to ensure that their invention is described correctly and precisely in the application documents, and in particular in the claims.[12] Deficiencies in the description are to the detriment of the applicant, since it was within their power to adequately formulate the technical teaching of the invention.

When interpreting claims, the disclosure content of the claims is the starting point, with the description and the drawings of the patent specification serving as aids to interpretation.[13] In particular, this can clarify ambiguous terms in a patent claim. In addition, the extent of the scope of protection can be determined.[14] The description and the drawings may only be understood as an explanation of the claims.[15] The patent claims define the subject matter protected.[16] In order for the patent claims to fulfill this function, they must be clearly formulated. If the examiner at the patent office is presented with unclear claims during the grant procedure, the patent application is to be rejected.[17]

The interpretation of a patent claim is a judicial task in patent infringement proceedings and leads to the legal determination of the scope of protection of a patent.[18]

Claims require interpretation in order to determine the scope of protection conferred by a patent's right to prohibit. Interpretation is always necessary, even if the wording of a claim may appear self-explanatory. This is already evident from the nature of a patent, as a patent constitutes its own lexicon. Terms in a patent may be defined differently from their usual meaning in order to appropriately describe the invention. For this reason, claims must always be interpreted in light of the entire disclosure of the patent. Proper

---

[10] Osterrieth, Part 6. Patent Infringement, Osterrieth, Patent Law, 6th edition 2021, paras. 856–860; Schulte/Rinken, PatG § 14 paras. 15 ff.

[11] BGHZ 100, 249 (254)—Broadcasting Transmission System.

[12] Osterrieth, Part 6. Patent Infringement, Osterrieth, Patent Law, 6th edition 2021, para. 859.

[13] Mes, 5th ed. 2020, PatG § 34 para. 31.

[14] BGH GRUR 2004, 413, 414, right column—Geflügelkörperhalterung; 2002, 511—Kunststoffrohrteil; BGHZ 150, 149, 153 = GRUR 2002, 515—Schneidmesser I; 2002, 519, 521, left column—Schneidmesser II; 2002, 523, 524—Custodiol I; 2002, 527, 528, 529—Custodiol II; BGHZ 125, 303, 309 = GRUR 1994, 597—Zerlegvorrichtung für Baumstämme; BGHZ 105, 1, 10 = GRUR 1988, 896—Ionenanalyse; BGHZ 98, 12, 18 = GRUR 1986, 803—Formstein.

[15] Osterrieth, Part 6. Patent Infringement, Osterrieth, Patent Law, 6th edition 2021, para. 866.

[16] § 14 sentence 1 Patent Act: "The scope of protection of the patent and the patent application is determined by the patent claims"; BGH GRUR 2000, 1005—Bratgeschirr.

[17] § 34 paragraph 3 no. 3 Patent Act.

[18] Nägerl, § 4. Distinction from the State of the Art—Novelty and Inventive Step, Haedicke/Timmann, Handbook of Patent Law, 2nd edition 2020, para. 46.

interpretation of the claims ensures adequate protection for the patent holder and legal certainty for third parties.[19]

The purpose of interpretation is, in particular, to eliminate ambiguities and to ensure the correct understanding of the terms used.[20] When interpreting claims, the starting point is the wording of the relevant claim.[21] However, for interpretation, it is mandatory to refer to the description and drawings of the patent.[22] The claims, the description, and the drawings are to be regarded as a unit in the process of interpretation. However, the primacy of the claim wording must be recognized.[23]

A legal norm in jurisprudence describes an abstract legal situation. Before the legal consequences of a legal norm can be enforced, it must be examined whether the legal norm is applicable to the specific legal situation. This is the purpose of interpretation. Interpretation examines whether the legal norm applies to the specific legal situation.[24]

In jurisprudence, interpretation can in particular be carried out according to four different variants. There is a literal, a systematic, a teleological, and a historical interpretation. All variants of interpretation are to be applied to a legal norm. These variants of interpretation developed in jurisprudence can, with the exception of the historical method, be applied to patent claims.

The variants of interpretation are not alternatives, but all must be applied to a patent claim in order to balance the tension between the appropriate scope of protection for the patent holder and legal certainty for the public.[25] In addition to the literal, systematic, and teleological interpretation, a function-oriented interpretation can be applied to patent claims. The interpretation of patent claims is carried out from the perspective of a person

[19] Article 1 of the Protocol on the Interpretation of Article 69 EPC as amended by the Act revising the EPC of 29 November 2000, https://www.epo.org/law-practice/legal-texts/html/epc/2020/d/ma2a.html, accessed on 1.9.2024.

[20] German Federal Court of Justice (BGH), judgment of 12 March 2002—X ZR 135/01, GRUR 2002, 519—Schneidmesser II.

[21] Section 14 sentence 1 Patent Act or Article 69(1) sentence 1 EPC.

[22] Section 14 sentence 2 Patent Act or Article 69(1) sentence 2 EPC: "Nevertheless, the description and the drawings shall be used to interpret the claims."

[23] Section 14 sentence 1 Patent Act.

[24] Reinhold Zippelius, Juristische Methodenlehre, 11th edition 2012, before § 1, § 10; Karl Engisch, Die Idee der Konkretisierung in Recht und Rechtswissenschaft unserer Zeit, 2nd edition, Heidelberg 1968; Karl Larenz, Methodenlehre, 1st edition 1960, Chapter 4; Bernd Rüthers, Methodenlehre, para. 698.

[25] Article 1 of the Protocol on the Interpretation of Article 69 EPC as amended by the Act revising the EPC of 29 November 2000, https://www.epo.org/law-practice/legal-texts/html/epc/2020/d/ma2a.html, accessed on 1.9.2022.

skilled in the art working in the relevant technical field. The person skilled in the art is a fictitious person.[26]

A historical interpretation would consider the genesis of the claim. In contrast to a legal norm, for which it is possible to determine what the legislator originally intended and what considerations guided them, the genesis of an invention as the subject of a claim cannot be reliably established. Therefore, historical interpretation of patent claims does not take place. Moreover, a subjective interpretation would lead to legal uncertainty and would be contrary to the absolute (objective) concept of novelty.[27]

4.1 Purpose of Interpretation

The purpose of interpreting a claim is to provide the applicant with appropriate protection and, on the other hand, to ensure sufficient legal certainty through a comprehensible, objective determination of the terminology used in the claim.

4.2 Main Claim, Parallel Claim, Dependent Claims, Independent and Dependent Claims

The main claim is the first claim of a set of claims and contains the essential features of the invention.[28] A set of claims may include claims in addition to the main claim that do not refer to any preceding claim. These claims are referred to as parallel or auxiliary claims.[29] Most of the time, the parallel claims belong to a different patent category than the main claim. All other claims are dependent claims, which refer to preceding claims and therefore include all features of the respective referenced claims.[30] The main claim and the parallel claims are referred to as independent claims.[31] The dependent claims are alternatively referred to as dependent claims.

Figure 4.1 shows the assignment of main claims, parallel claims, and dependent claims to independent and dependent claims.

[26] BGH, judgment of 9 January 2018—X ZR 14/16, GRUR 2018, 390—Wärmeenergieverwaltung; BGH, judgment of 12 March 2002—X ZR 43/01, GRUR 2002, 511—Kunststoffrohrteil.

[27] Benkard EPC/Scharen, 4th ed. 2023, EPC Art. 69 para. 15.

[28] Section 9(4) Patent Ordinance.

[29] Section 9(5) sentence 1 Patent Ordinance.

[30] Section 9(6) sentence 1 Patent Ordinance.

[31] Section 9(5) sentence 1 Patent Ordinance.

Fig. 4.1 Types of claims

independent claims	Main claim
	Secondary claim
Dependent claims	Subclaims

Product claim	Device claim	Machine, device, work equipment
	Substance claim	Chemical material
Process claim	Manufacturing claim	Steps for manufacturing a product Analyses, evaluations, processing
	Application claim	Steps for the correct application of a substance or device

Fig. 4.2 Categories of claims

4.3 Categories of Claims

Essentially, product claims and process claims can be distinguished, with product claims being subdivided into apparatus claims and substance claims. A process claim can be a manufacturing claim or a use claim. Figure 4.2 provides an overview of the different patent categories with examples.

4.4 Literal Interpretation

In literal or word-for-word interpretation, the meaning of each term in the claim is determined. The sum of the terms yields the overall meaning of the claim. A literal interpretation is only the starting point for interpreting a patent claim. At least a systematic interpretation is also mandatorily required.[32]

A claim must be clear and unambiguous in itself.[33] However, this clarity is only required from the perspective of a person skilled in the art, who uses their technical understanding along with their specialized expertise and skills to extract the meaning of the claims.[34]

[32] Section 14 sentence 2 Patent Act.

[33] Benkard EPC/Wieser/Kinkeldey, 4th ed. 2023, EPC Art. 84 para. 15.

[34] Benkard EPC/Wieser/Kinkeldey, 4th ed. 2023, EPC Art. 84 para. 20.

When interpreting claims, every word of the claim must be considered. However, interpretation should not be rigidly bound to the wording. The words of the claim are therefore not to be understood literally, but are to be classified from the perspective of the hypothetical skilled person.[35]

Literal interpretation forms the starting point for any interpretation of claims.[36] Here, the meaning of the claims is to be determined based on their wording in the context of the description and the drawings.[37] Thus, interpretation of a claim always also involves a systematic interpretation, whereby the meaning is determined from the context of the description, the drawings, and the remaining claims of the claim set.

4.5 Systematic Interpretation

The systematic interpretation of a claim determines the meaning of a claim from the context of the description, the drawings, and the remaining claims. Systematic interpretation serves in particular to clearly define ambiguous terms in the claim or to resolve contradictory formulations.[38]

A discrepancy between the literal and systematic interpretation of the claims may arise if the description and drawings are at odds with the wording of the claims. In such a case, the meaning derived from the wording of the claims prevails.[39]

A patent specification is considered its own dictionary, so the meaning of terms in the claims may deviate from their usual meaning. It is therefore necessary to consider the entire patent specification and, if required by the disclosure of the patent specification, to deviate from usual meanings.[40]

4.6 Teleological Interpretation

In teleological interpretation, the focus is on the sense and purpose expressed in the claim wording. In teleological interpretation, the technical problem is particularly used to understand the claim. Teleological interpretation can alternatively be referred to as problem-oriented interpretation.[41] Teleological interpretation of claims is viewed

[35] Benkard EPC/Scharen, 4th ed. 2023, EPC Art. 69 para. 15.

[36] Osterrieth, Part 6. Patent Infringement Osterrieth, Patent Law, 6th edition 2021, para. 861.

[37] Section 14 Patent Act.

[38] Benkard PatG/Schacht, 12th ed. 2023, PatG § 34 para. 162.

[39] Benkard EPC/Wieser/Kinkeldey, 4th ed. 2023, EPC Art. 84 para. 12.

[40] Benkard EPC/Wieser/Kinkeldey, 4th ed. 2023, EPC Art. 84 para. 13.

[41] Federal Court of Justice, judgment of 4.2.2010—Xa ZR 36/08, GRUR 2010, 602—Joint Arrangement.

critically, as what the applicant "actually" intended to achieve with the invention is often not clearly discernible from the description. Moreover, during the course of the patent granting procedure, the problem to be solved may change. Teleological interpretation is therefore fraught with a high degree of uncertainty.[42]

4.7 Function-Oriented Interpretation

Claims are to be interpreted from the perspective of the skilled person. The skilled person will seek the technical meaning, the intended function of the subject matter of the claim, and will determine this based on the state of the art, the technical problem posed, and the functions performed by the claimed embodiments.[43] The interpretation of the claims requires the comprehensive technical knowledge and usual skills of the skilled person to determine the correct scope of protection of the claims. By this approach, the skilled person can exclude embodiments that are technically nonsensical.[44]

In function-oriented interpretation, care must be taken to avoid impermissible generalization. Only those embodiments can be claimed that can be directly and unambiguously derived from the claim wording. The risk with function-oriented interpretation is that an understanding is adopted that did not exist on the filing or priority date and thus extends beyond the scope of protection.[45]

4.8 The Skilled Person as the Standard for Interpretation

The scope of protection resulting from the claims is determined from the perspective of the skilled person.[46] Guidance on how claims are to be interpreted can be found in the Protocol on the Interpretation of Article 69 EPC, which, pursuant to Article 164(1) EPC, forms part of the European Patent Convention. Accordingly, the scope of protection is not determined solely by the literal wording of the claims. In addition, it must be considered what the skilled person, in light of the description and drawings, understands as the

[42] Osterrieth, Part 6. Patent Infringement in Osterrieth, Patent Law, 6th edition 2021, para. 859.

[43] Timman, § 3. Interpretation and Scope of Protection of Patents, Haedicke/Timmann, Handbook of Patent Law, 2nd edition 2020, para. 51.

[44] Timman, § 3. Interpretation and Scope of Protection of Patents, Haedicke/Timmann, Handbook of Patent Law, 2nd edition 2020, para. 52.

[45] Benkard PatG/Scharen, 12th ed. 2023, PatG § 14 para. 71; Meier-Beck GRUR 2003, 905, 907.

[46] Osterrieth: The Skilled Person in Patent Law, GRUR 2021, 310, 312.

subject matter for which protection is sought.[47] Furthermore, the technical understanding of the skilled person is required to fill in any gaps.[48]

The interpretation of patent claims is a demanding undertaking that often requires balancing the potentially differing results of literal, systematic, teleological, and function-oriented interpretation—an endeavor that can only be properly addressed from the perspective of the skilled person.

Inventive step is to be assessed from the viewpoint of the knowledge and abilities of the skilled person as of the filing or priority date.[49] This necessarily follows from the fact that only the state of the art prior to the filing or priority date may be taken into account. Therefore, only the knowledge and abilities of the skilled person available before the filing or priority date are to be considered.[50]

[47] Osterrieth: The Skilled Person in Patent Law, GRUR 2021, 310, 312.

[48] Benkard PatG/Scharen, 12th ed. 2023, PatG § 14 para. 25.

[49] BGH GRUR-RS 2021, 12752 para. 35—Dental Implant; EPO GRUR-Int 1983, 650, 651—Metal Finishing; EBA EPO G2/21 of 23.3.2023, Official Journal EPO 2023, A85.

[50] BGH GRUR 1969, 271, 272—Cable Guide; BGH GRUR 2004, 411, 413—Diabehältnis; BGH GRUR 1980, 899—Sourdough.

The Inventive Step

5

Contents

A skilled person does not apply the technical teaching of a document exactly as described. Instead, they adapt it according to the specific circumstances and their own expertise and skills. Even in quite ordinary activities, modifications arise, which the technician arrives at through the application of their prior knowledge and experience. These modifications, which routinely occur to the skilled person, do not deserve legal protection. Instead, there should be legal leeway here so as not to hinder the usual professional activity of a skilled person. If such modifications were already patentable,

T. H. Meitinger, *Justification of the Inventive Activity*,
https://doi.org/10.1007/978-3-662-72735-5_5

patent law would hinder the application of known technology.[1] The patent law skilled person is therefore needed to determine what still constitutes routine, technical skill of the average skilled person and thus does not represent a patentable inventive activity.

The criterion of inventive step is a qualitative criterion that assesses the creative achievement and creativity. A value judgment must be made.[2] The assessment of inventive step is one of the most difficult tasks in patent law. Skillful argumentation by the inventor can lead to the granting of a patent or a successful defense in contentious proceedings, whereas a less suitable justification can mean the loss of the proceedings or the patent.

The criterion of inventive step is an objective criterion. It does not depend on a subjective opinion, possibly of the applicant or the inventor.[3] It is not relevant what prior knowledge the inventor had when creating the invention. The inventor may have been mistaken in creating the invention, i.e., may not have understood at all why the invention produces the advantageous effect.[4] On the other hand, a lack of knowledge by the inventor about the state of the art does not lead to the inventive step of their invention.

The assessment of inventive step was previously regarded as a purely factual question, for which only the correct determination of the actual circumstances was required.[5] Since 1978, when the criterion of inventive step was incorporated into the Patent Act, it has become a legal question, requiring an interpretation of the relevant legal norm and a subsumption. Therefore, a legally evaluative assessment of the facts is required in order to assess inventive step.[6]

Inventive step is an indeterminate legal concept for which there is only one correct decision. It is a legal question and not a question of fact.[7] There is no discretion.

Three aspects must be considered when assessing inventive step:

- State of the art,
- Skilled person, and
- Obviousness.

---

[1] Ann, § 18. Inventive Step in Ann, Patent Law, 8th ed. 2022, para. 2.

[2] BGH GRUR 1995, 330, 331—electrical plug connection; BGH GRUR 2004, 411, 413—diabetes ratio; BGH GRUR 2010, 410—occupant protection system control unit.

[3] BGH GRUR 1987, 510—middle ear prosthesis.

[4] BGH GRUR 1965, 138, 142—polymerization accelerator; BGH GRUR 1994, 357, 358—muffle furnace.

[5] BGH GRUR 1984, 797—tine rotor; BGH GRUR 1987, 510—middle ear prosthesis; BGH GRUR 1998, 913—induction furnace.

[6] BGH GRUR 95, 330 electrical plug connection; BGB GRUR 04, 411 diabetes ratio; BGH GRUR 06, 663 prepaid telephone calls.

[7] BGH GRUR 06, 663 prepaid telephone calls; BGH GRUR 06, 842 demonstration cabinet.

An assessment of inventive step is only possible if the relevant state of the art and the applicable skilled person are first determined. The relevant skilled person is the one at the time of the filing or priority date. The state of the art includes all disclosures made available to the public before the filing or priority date. The hypothetical patent law skilled person will therefore assess inventive step with the knowledge available before the filing or priority date.[8]

Since the question of inventive step is not a question of fact but a legal question, the assessment of inventive step cannot be made by a technical expert.

An invention can only have sufficient inventive height if it demonstrates inventive step over every document of the state of the art and every combination of documents from the state of the art that would make sense to the skilled person. If a lack of inventive step is demonstrated by a combination of documents from the state of the art, further examinations, possibly with another closest state of the art, can be omitted, as a change in the assessment is no longer possible.[9]

It is not required that the invention represents an improvement or a technical advance over the state of the art. It may be sufficient if an alternative solution to a technical problem is proposed.[10]

5.1 Indeterminate Legal Concept of "Inventive Step"

In assessing inventive step, a value judgment of a legal question is involved, not a question of fact. The question of inventive height therefore cannot be decided by a technical expert. A technical specialist can determine whether 100 V or 200 V is present at the output of a circuit. However, this does not answer the question of inventive step.

The question of inventive step requires a value judgment, for example, to determine whether the production of a superconducting material at −200 °C or −220 °C is inventive if superconductivity at −240 °C is already known. Furthermore, if a claim specifies 100 V as a feature, the question may arise as to whether a circuit that generates 110 V still infringes the claim. Such questions will not be answered by a technical expert.[11]

In patent law, inventive step is an indeterminate legal concept, as the Patent Act, like the legal concept of invention, does not provide a definition. The determination of inventive step is the responsibility of the legal practice of the patent offices and the case law of the courts, in particular the Federal Patent Court, the patent litigation chambers of the regional courts, and the Federal Court of Justice.

[8] EPO T 24/81.

[9] EPO T 1742/12.

[10] EPO T 1791/08.

[11] Beckmann, On the Differentiation and Quantification of Inventive Height, Infringement of Intellectual Property Rights, and Legal Consequences in Patent Law, GRUR 1998, 7, 8.

5.2 Purpose of the Inventive Step Criterion

The criterion of inventive step is intended to prevent a thicket of patents that would unduly burden the activities of the skilled person in their technical field. In particular, the European Patent Office, with its "Raising the bar" initiative, aimed to prevent so-called trivial patents.[12]

A skilled person never applies a technical teaching from the prior art exactly as described. The skilled person always incorporates at least their expertise when applying it, so that a (minor) modification of the technical teaching is regularly implemented.[13] If such a minor variation could already be monopolized by a patent, patent law would hinder the technical application of known knowledge. Obvious modifications of a technical teaching therefore cannot be claimed by a patent.

Patents are granted only for special achievements and serve as an incentive to enrich technical knowledge.[14] Patents should therefore not be granted for trivialities. This results in an intermediate area around the prior art that is in the public domain.[15] One purpose of the inventive step criterion is therefore to avoid trivial patents that would hinder economic and technological development.[16] If the invention is outside this public domain area, the invention demonstrates sufficient inventive step for patent grant.

An invention is therefore only patentable if it stands out from what is to be expected and thus maintains a sufficient distance from the prior art. It is obvious that the criterion of inventive step is considerably more difficult to apply compared to the requirement of novelty.

5.3 Inventive Step of a Utility Model

The Utility Model Act does not require "inventive activity" (erfinderische Tätigkeit) for a legally valid utility model, but rather an "inventive step" (erfinderischer Schritt). The intention of the legislator was to create, with the Utility Model Act, a protective right that requires a lower threshold compared to patent law.[17]

---

[12] Koch, The Feature of Inventive Step as a Corrective in Patent Law, GRUR Int 2008, 669.

[13] Nähring, GRUR 1959, 57 f.

[14] BGH GRUR 87, 231; 96, 109 Clinical Trials I; 97, 454 Cable Feedthrough; Mitteilungen der deutschen Patentanwälte 97, 253 Clinical Trials II.

[15] BGH GRUR 06, 842 Demonstration Cabinet; Goebel GRUR 08, 301; Keukenschrijver GRUR Int 08, 665; Wenzel GRUR 13, 140.

[16] Koch, The Feature of Inventive Step as a Corrective in Patent Law, GRUR-Int 2008, 669, 670.

[17] BPatG Mitteilungen der deutschen Patentanwälte 2002, 46—Cable Duct; BPatG GRUR 204, 852—Material Strip Packaging; BPatG GRUR 2006, 489—Tool for Joining Profile Boards; BPatG BeckRS 2012, 19.103—Drilling Device; BPatG BIPMZ 2006, 36—Toilet Seat Riser.

After many years in which the difficulties in handling an inventive step with a lower inventive threshold became increasingly apparent, the case law no longer followed the legislator's intention and decided in the Demonstration Cabinet decision[18] that the same requirements must be applied to the inventive step[19] of a utility model as to the inventive step[20] of a patent.

5.4 Objective Assessment of Inventive Step

An invention is examined for its objective inventive step. It is irrelevant whether it was difficult or easy for the present inventor to arrive at the invention. Whether the inventor is a Nobel laureate or a cleaner without technical training is not taken into account. The individual effort required to create the invention is of no importance. The inventive step is to be assessed from the perspective of an average skilled person. Only if the distance between the prior art and the invention is so great that the invention is no longer obvious to the patent law skilled person is there sufficient inventive step for a patent grant.[21]

It would be wrong to regard the granting of a patent as a badge of honor for an invention because of its particular creative achievement. In fact, the criterion of inventive step is already met for any performance that is not merely average.[22]

5.5 Non-Technical Features

According to Section 1(1) Patent Act, patents are granted only for inventions in the technical field. If an invention contains exclusively non-technical features, patenting is excluded. If a claim contains both technical and non-technical features, the non-technical features cannot establish inventive step.[23] This does not apply if the inventive effect

[18] BGH, Inventive Step in Utility Model Law—Demonstration Cabinet—NJW, 2006, 3208 and GRUR 2006, 842, 845—Demonstration Cabinet.

[19] Section 1(1) Utility Model Act.

[20] Section 4 Sentence 1 Patent Act.

[21] Section 4 Sentence 1 Patent Act.

[22] BGH, Inventive Step in Utility Model Law—Demonstration Cabinet—NJW, 2006, 3208 and GRUR 2006, 842, 845—Demonstration Cabinet.

[23] EPO 8.9.2000—T 931/95, OJ EPO 2001, 441; EPO 4.4.2000—T 158/97, BeckRS 2000 30529657; EPO 9.7.2002—T 1177/97, BeckRS 2002 30684643; EPO 26.9.2002—T 641/00, OJ EPO 2003, 352; EPO 12.7.2005—T 914/02, BeckRS 2005 30657298; EPO 22.3.2006—T 619/02, OJ EPO 2007, 63; EPO 17.3.2005—T 531/03, BeckRS 2005 30600453; EPO 7.3.2007—T 1284/04, BeckRS 2007 30689156; EPO 27.10.2006—T309/05, BeckRS 2006 30562011; EPO 15.4.2008—T 812/05, BeckRS 2008 30657028; EPO 23.6.2010—T 784/06, Special Edition OJ EPO 2011, 39 f.; EPO 2.9.2008—T 1814/07, BeckRS 2008, 145852.

only arises through the interaction of the non-technical with the technical features. In this case, the non-technical features must also be taken into account when assessing inventive step.[24]

As a rule, however, all non-technical features are disregarded when assessing inventive step[25] and only those features that have a technical character and serve to solve the technical problem are taken as a basis for the assessment of inventive step.[26] The mere transmission of knowledge or features that are purely aesthetic in nature cannot ensure inventive step. The same applies to technically nonsensical features.[27]

5.6 State of the Art

The state of the art encompasses all documents, disclosures, presentations, and any other form of existing know-how that was made available to the public before the filing or priority date of the patent application.[28]

It is assumed that, at the priority date—the filing or priority date—the skilled person had access to the complete state of the art. It is irrelevant how difficult it was to identify the state of the art. An exception may arise if public access was virtually excluded due to prohibitively high access costs.[29]

In contrast to the novelty examination, which involves a one-to-one comparison, the state of the art is combined in a mosaic-like manner by the skilled person when assessing inventive step.[30] The disclosure content of the state of the art corresponds to what the skilled person would derive from it.[31]

If the skilled person has reason to use a document and it describes several alternative embodiments, the skilled person will consider each of these embodiments.[32]

---

[24] EPO 29.7.1983—T37/82, OJ EPO 1982, 71; EPO 15.11.2006—T154/04, OJ EPO 2008, 46; EPO 13.12.2006—T1227/05, OJ EPO 2007, 574; EPO T1171/06; EPO 27.1.2010—T354/07, BeckRS 2010, 146235; EPO 11.7.2013—T1670/07, Supplementary Publication OJ EPO 2014, 41.

[25] EPO 8.9.2000—T931/95, OJ EPO 2001, 441; EPO 26.9.2002—T641/00. OJ EPO 2003, 352; EPO 27.11.2003—T172/03, BeckRS 2003 30533185; EPO 24.11.2003—T1121/02, beckRS 2003 3068046; EPO 24.11.2003—T258/03, OJ EPO 2004, 575; EPO 18.4.2008—T756/06, BeckRS 2008, 145275; EPO 21.9.2012—T1784/06, BeckRS 2012, 213594; EPO 13.10.2025—T483/11, Supplementary Publication OJ EPO 2017, 22.

[26] BGH GRUR 2020, 599 para. 24—Rotating Menu.

[27] BGH GRUR 2015, 983 para. 31—Aircraft Condition.

[28] Section 3, paragraph 1, sentence 2 of the Patent Act.

[29] Federal Court of Justice (BGH) Bausch 1994/1999, 159, 162—Concrete Rings; EPO GRUR-Int 1986, 545, 546—Boeing; BGH Liedl 1963/1964, 157, 166.

[30] Benkard PatG/Asendorf/Schmidt/Tochtermann, 12th ed. 2023, PatG § 4 para. 33.

[31] BGH—Pipe Clamp; BGH GRUR 54,24—Multiple Clamp; BGH GRUR 64, 167—Writing Instrument; BGH GRUR 74, 208—Power Supply Device.

[32] EPO 11.10.2019—T 787/17, Special Edition 2 OJ EPO 2021, 12.

If two documents from the state of the art contradict each other, the skilled person will exclude a combination of these documents.

5.7 Closest Prior Art

The closest prior art is the state of the art that the skilled person would use as a starting point to arrive at the technical teaching of the invention under examination.

The European Patent Office considers as the closest prior art the most promising document for arriving at the invention. In European proceedings, the document is regarded as closest prior art if it pursues the same purpose or objective, intends the same effect, and shares the essential features with the invention under examination.[33] The closest document should therefore exhibit the fewest structural differences from the invention. The most important criterion in selecting the closest prior art is the similarity of the technical problem. The number of identical features is less significant. The closest document in the state of the art should therefore have the same or a similar purpose or aim for the same effect.[34] Furthermore, the closest prior art should belong to the same or at least an adjacent technical field as the invention under examination.

It is entirely possible for several documents to be regarded as the respective closest prior art. In this case, the inventive step of the invention must be assessed starting from each individual document.[35]

However, as a rule, only one document is to be regarded as the closest prior art. In exceptional cases, several documents may be accepted together as the closest prior art. For example, if several patent specifications originate from the same inventor and are also related in content, the technical teachings of several patent specifications may, by way of exception, be used together as the closest prior art.[36]

If a document does not disclose a problem that at least resembles the objective technical problem leading to the invention, that document should generally not be used as the

---

[33] EPO 18.9.1990—T 606/89 OJ EPO 1991, 24; EPO 26.11.1993—T 570/91, OJ EPO 1995, 35 f.; EPO 16.5.1995—T 1040/93, BeckRS 1995 30672905; EPO 5.2.1997—T 506/95, OJ EPO 1998, 26; EPO 9.2.2017—T 1747/12, BeckRS 2017, 116208; EPO T 2255/10.

[34] EPO T 1148/15; EPO T 698/10.

[35] EPO 25.10.2001—T 967/97, BeckRS 2001 30664708; EPO 18.2.2004—T 558/00, BeckRS 2004 30604708; EPO 2.9.2010—T 21/08 BeckRS 2010, 146886; EPO 8.2.2012—T 1289/09, BeckRS 2012, 215691; EPO 22.6.2016—T 1742/12 Beck RS 2016, 120569; EPO 28.2.2014—T 1437/09 Special Edition 4 OJ EPO 2015, 34; EPO 9.2.2011—T 308/09 BeckRS 2011, 146156; Examination Guidelines 2021 G VII, 5.1; EPO 23.3.2017—T 1570/13, BeckRS 2017, 122845; EPO 10.1.2018—T 855/15, BeckRS 2018, 1242.

[36] EPO 27.6.1990—T 176/89, RprBK 2019, 205; EPO 3.11.2010—T 53/08 Rspr. BK 2019, 207; EPO 11.1.2018 -T 2571/12 Rspr. BK 2019, 207.

closest prior art. In such a case, the skilled person would not consider the document to solve the objective technical problem. It is therefore not suitable as the closest prior art.

The use of a closest prior art can be regarded as a procedural economy approach. It can be argued that if an inventive step is present starting from the closest prior art, it will also be present when starting from any other document in the state of the art.[37]

It is not mandatory to define a closest prior art. In particular, the determination of a closest prior art may carry the risk of an inadmissible ex post facto analysis. In principle, the assessment of inventive step should be carried out starting from each document in the state of the art.[38]

The selection of a document as the closest prior art must be justified. As a rule, the justification is that the skilled person is seeking a better or alternative solution to the existing state of the art.[39]

If a revolutionary, groundbreaking invention is to be assessed, it is very difficult to determine a meaningful closest prior art.

In proceedings before the Unified Patent Court, a realistic starting point must be sought to solve the technical problem. There may be several realistic starting points. From the perspective of the Unified Patent Court, a starting point is realistic if the document solves a similar or the same problem, or if a similar product or process is developed with the document.

For the European Patent Office, the closest prior art is, in particular, the document that requires the fewest structural modifications to arrive at the invention under examination.

5.8 Subsequently Published Prior Art

Subsequently published prior art refers to patents or patent applications that were filed with the patent office before the filing or priority date of a patent application to be examined for patentability, but which were not published until on or after the filing or priority date of the application.[40] These patent documents can call into question the novelty of a patent application[41], but they are not taken into account when assessing inventive step.[42] This is appropriate, because during the period in which the patent office kept the patent

---

[37] EPO T 816/16.

[38] Jestaedt GRUR 2001, 939, 941; EPO GRUR-Int 1996, 723—Aluminium Alloy/ALCAN; EPO Mitteilungen der deutschen Patentanwälte 2002, 315—Chip Card.

[39] BGH GRUR 2009, 382 para. 51—Olanzapine; BGH GRUR 2017, 148 paras. 42 f.—Optoelectronic Component; BGH GRUR 2017, 498, para. 28—Knitted Shoe Upper; BGH GRUR-RS 2020, 38.340 para. 33.

[40] Section 3(2) Patent Act; Article 54(3) EPC.

[41] Section 3(2) Patent Act; Article 54(3) EPC.

[42] Section 4, sentence 2 Patent Act; Article 56, sentence 2 EPC.

document confidential, it could not have been combined with other documents to arrive at the invention in an obvious manner. Subsequently published prior art is therefore irrelevant when assessing inventive step. The consideration of subsequently published prior art in the novelty examination is intended to prevent double patenting of the same invention.

5.9 Technical Problem

The assessment of inventive step is generally not based on the problem that the inventor set out to solve at the time of creating the invention. The inventor's subjective approach is disregarded. Instead, the focus is on the objective technical problem that arises from the prior art available on the relevant date. The objective technical problem is directly and immediately related to the advantageous effects of the invention.[43]

The technical problem is the objective issue that the skilled person solves with the invention. The problem is to be determined in the examination proceedings before the patent office on the basis of the prior art documents. Reformulation of the technical problem is permissible at any stage of the examination proceedings and even during the applicant's appeal proceedings.[44]

It is entirely possible that the inventor arrived at the invention only thanks to a particularly ingenious formulation of the problem. In this sense, the problem conceived by the inventor forms part of the invention. However, for the assessment of inventive step, an objective technical problem must be determined, which the hypothetical patent law skilled person derives from the prior art encountered.[45] The objective technical problem is the starting point for assessing whether an invention is obvious or not.[46] If new prior art is identified, it may be necessary to reformulate the objective technical problem.[47]

The problem must have a technical character.[48] In addition, the problem must be directed to effects that are directly and causally related to the features of the invention under examination.[49] The problem must be formulated in such a way that it does not suggest any solutions.[50]

---

[43] EPO Enlarged Board of Appeal G2/21, Official Journal 2023, A85.

[44] BeckOK PatR/Einsele, 33rd ed. 15.7.2024, EPC Art. 56 paras. 4a-4b.

[45] German Federal Court of Justice (BGH) GRUR 91, 522—Fire Protection Closure; BGH GRUR 1987, 510—Middle Ear Prosthesis.

[46] BGH BeckRS 2016, 13031—Pemetrexed.

[47] EPO T GRUR-Int 84, 525; T 13/84 OJ 1986, 253; EPO T 162/85 OJ 1988, 452; EPO TK GRUR-Int 1994, 516.

[48] EPO G 1/19.

[49] EPO G 2/21.

[50] EPO T 229/85; EPO T 1252/14.

It is permissible to reformulate the technical problem.[51] However, the new problem must at least be derivable from the present application documents.[52]

The technical problem must be formulated in a neutral manner, so that the wording of the problem does not already provide an indication of the solution. In particular, the formulation of the problem must not render the invention obvious.[53]

In proceedings before the German Patent Office, the emergence of new prior art does not necessarily require the technical problem to be changed.[54]

5.10 Prohibition of Ex Post Facto Analysis

An ex post facto approach is not permitted. Instead, the assessment of inventive step must be made from the perspective of the skilled person on the filing or priority date. It should be noted that the skilled person had access to the technical knowledge and prior art as of the day before the filing or priority date, i.e., at a time when the patent application had not yet been filed. Technical knowledge or expertise acquired by the skilled person on or after the filing or priority date cannot be taken into account in the assessment.[55]

It is questionable whether special effects, advantages, or results arising from a technical teaching may be taken into account in the assessment of inventive step if such special effects, advantages, and results are not disclosed in the prior art. However, if the invention is specifically aimed at these special effects, they may not be considered as part of the prior art in the assessment. Otherwise, the prior art is to be used by the skilled person as it is presented to them, in order to arrive at the invention.[56]

It is permissible to use publications that only became available after the filing or priority date to aid in understanding the content of documents that were published before the filing or priority date.[57]

[51] EPO T 1397/08.

[52] EPO T 13/84; EPO T 877/06; EPO G 2/21.

[53] BGH GRUR 2024, 1432—Mirabegron.

[54] BGH GRUR 1991, 811—Folding Machine.

[55] BGH GRUR 1980, 100, 103—Floor Sweeper; BGH GRUR 1981, 338—Magnetic Field Compensation; BGH GRUR 1989, 899, 902—Sourdough; BGH GRUR 2001, 232, 234—Hole Punch; BGH Mitteilungen der deutschen Patentanwälte, 2003, 116, 120—Stirrer; BGH GRUR 2019, 925 para. 18—Bitrate Reduction II.

[56] Benkard PatG/Asendorf/Schmidt/Tochtermann, 12th ed. 2023, PatG § 4 para. 32; BGH GRUR 1962, 83, 85—Insole; BGH GRUR 1971, 403, 406—Pallet Truck; BGH BeckRS 2018, 40825 para. 46; BGH GRUR-RS 2021, 4292 para. 135.

[57] BGH GRUR 1964, 612, 616—Beer Bottling.

5.11 Obviousness

According to § 4 sentence 1 of the Patent Act and Article 56 sentence 1 EPC, an invention is based on inventive step only if the invention is not obvious to a person skilled in the art in view of the prior art. The "normal" further development of technology is therefore excluded from patent protection.

If a technical teaching logically follows from the prior art for a person skilled in the art, the criterion of obviousness is met and a patent grant is excluded. However, the skilled person must also have had a motivation to seek a solution to the technical problem in the direction that leads to the technical teaching in question.[58]

If an invention is obvious, it is not based on inventive step.[59] If a technical teaching can be achieved by routine expertise based on the knowledge of a person skilled in the art, the technical teaching is obvious to the skilled person. If an invention is a necessary or at least logical consequence of the known prior art, then no technical progress can have been made by the invention and the technical teaching of the invention is obvious. However, obviousness is not limited to what is immediately apparent. The intermediate area between the prior art and the field of inventive step encompasses everything accessible to the knowledge and skills of the average skilled person, without being creative.[60]

Obviousness means that the invention was apparent to the skilled person, or that there were sufficient prompts, suggestions, or indications for the skilled person to take a technical path leading to the invention.[61]

A sufficient reason for a skilled person to pursue a particular solution path may already arise from laws or legal regulations.[62]

The mere avoidance of disadvantages, the optimization of parameters, and digitization or automation do not in themselves constitute inventive step, but rather represent a general trend in technology.[63]

A change in process steps for manufacturing a product is obvious if the skilled person arrives at the technical solution through usual considerations.[64] An invention must therefore go beyond normal technical progress in order to demonstrate inventive step. Sufficient inventive height can only be present if normal technical development is surpassed.[65] This means that a patent can only be granted if the teaching of the invention accelerates the development of technology beyond the usual level.

---

[58] BeckOK PatR/Einsele, 33rd ed. 15.7.2024, EPC Article 56 para. 8.

[59] § 4 sentence 1 Patent Act.

[60] Benkard EPC/Söldenwagner, 4th ed. 2023, EPC Article 56 para. 71.

[61] BGH GRUR 2009, 746 para. 20—Operation of a safety device.

[62] BGH judgment of 10 December 2013, X ZR 4/11, GRUR 2014, 349—Anthocyan compound.

[63] EPO 24.6.1992—T775/90; EPO 16.3.2005—T1175/02; EPO T 85/83 OJ EPO 1998, 183.

[64] EPO 4.5.1981—T1/81, OJ EPO 1981, 439.

[65] Benkard EPC/Söldenwagner, 4th ed. 2023, EPC Article 56 para. 71.

A patentable enrichment of technology may consist in solving a previously unknown problem, solving a previously unsolved problem, solving an already solved problem in an improved manner, or providing an alternative solution for an already solved problem.[66]

If a technical teaching belongs to the general knowledge of the skilled person, this does not automatically mean that it is obvious for the skilled person to apply this technical teaching.[67] However, the skilled person is motivated to draw on their general knowledge to solve a problem if the application of general knowledge is objectively expedient and not associated with significant difficulties.[68]

In assessing obviousness, it may be necessary to consider a variety of aspects. The assessment of obviousness can therefore be a complex process that must be carried out with the aid of auxiliary considerations.

Obviousness is particularly present if the skilled person had a reason to combine the relevant documents. The skilled person may have sufficient motivation to combine certain documents, especially due to an expected improvement or advantage, i.e., an anticipated success. When considering the expectation of success, it should be noted that in a technical field that has been little researched, predicting technical success is significantly more difficult compared to a well-researched field. An expectation of success can be assumed if, based on their knowledge, the skilled person could reasonably expect a corresponding success within a reasonable time before the invention was made.[69] An invention situated in an unknown technical field is therefore more likely to be inventive in the sense of patent law.

For the Unified Patent Court, an invention is considered obvious if, starting from a realistic closest prior art, the skilled person had a reason to consider a second document and thereby arrived at the invention under examination. A realistic closest prior art in particular addresses the same problem or purpose. Several documents may be regarded as realistic closest prior art, so that the examination scheme must be run through several times.

---

[66] Benkard EPC/Söldenwagner, 4th ed. 2023, EPC Article 56 para. 72.

[67] BGH GRUR 2009, 743 para. 37—Airbag release control; BGH GRUR 2018, 716 para. 28—Children's bed; BGH GRUR 2023, 39 para. 88—Light-emitting diode.

[68] BGH judgment of 11 March 2014—X ZR 139/10, GRUR 2014, 647—Ink supply system; BGH GRUR 2018, 716—Children's bed; BGH GRUR 2021, 1277—Guide rail arrangement; BGH judgment of 3 September 2024—X ZR 106/22—Disc brake III.

[69] Benkard EPC/Söldenwagner, 4th ed. 2023, EPC Article 56 para. 78.

5.12 Could-Would Test

The European Patent Office assesses obviousness using the could-would test.[70] According to the could-would test, an invention is not already obvious simply because a combination of prior art documents discloses the invention. Rather, the skilled person must also have had a reason to combine the relevant documents. In particular, the skilled person has a reason to combine documents if the documents themselves provide indications that a combination is possible and makes sense.[71]

The skilled person in patent law is therefore particularly necessary in the context of the could-would test. This test makes it possible, when examining inventive step, to consider only those documents that a skilled person would regard as combinable.[72]

The could-would test is based on the assumption that a skilled person does not act without reason. Without a specific reason, the skilled person will not combine documents.[73]

The could-would condition is met, and thus the relevant documents are combinable, if the skilled person would consult a document in order to expand the technical teaching of the closest prior art. The skilled person must therefore have an indication, a reason, or a suggestion to combine various prior art documents.[74]

---

[70] BeckOK PatR/Einsele, 33rd ed. 15.7.2024, EPC Article 56 para. 3.

[71] EPO T1/80 OJ 1981, 206; EPO T20/81 OJ 1982, 217; EPO T 24/81 OJ 1983, 133; EPO T 2/83 OJ 1984, 265; EPO T248/85 OJ 1986, 261; EPO T 254/86 OJ 1989, 115; EPO T 7/86 OJ 1988, 381; EPO T 939/92 OJ 1996, 309; EPO T 61/90 OJ 1994 Special Edition 39; EPO T 203/93 OJ 1995, Special Edition 43; EPO T 597/92 OJ 1996, 135; EPO T 167/93 OJ 1997, 229; EPO T 939/92 OJ 1996, 309; EPO T 967/97 OJ Special Edition 3, 2003, 21; BGH BeckRS 2010, 03881; BGH GRUR 2006, 930—Microtome.

[72] Benkard PatG/Asendorf/Schmidt/Tochtermann, 12th ed. 2023, PatG § 4 para. 86.

[73] EPO 15.3.1984—T 2/83, OJ EPO 1984, 265; EPO 2.4.1985—T90/84, EPOR 79–85 C952; EPO 15.10.1985—T223/84, EPOR 1986, 66; EPO 16.1.1986—T124/84, EPOR 1986, 297; EPO 17.12.1986—T265/84, EPOR 1987, 193; EPO 1.2.1988—T392/86, BeckRS 1988 30576874; EPO 11.3.1988—T219/87, BeckRS 1988 30543376; EPO 28.4.1988—T 274/87, EPOR 1989, 207; EPO 16.9.1987—T7/86, OJ EPO 1988, 381; EPO 19.12.1991—T 513/90, OJ EPO 1994, 154; EPO 10.2.1993—T 564/89, OJ EPO 1993, 39; EPO 22.6.1993—T 61/90, OJ EPO 1993, 39; EPO 1.9.1994—T 203/93, OJ EPO 1994, 42 f.; EPO 1.3.1995—T 597/92, OJ EPO 1996, 135; EPO 3.5.1996—T 167/93, OJ EPO 1997, 229; EPO 22.6.1982—T 22/82, OJ EPO 1982, 341; BGH 30.4.2009—Xa ZR 92/05, GRUR 2009, 746 para. 18 ff.—Operation of a safety device; BGH 8.12.2009—X ZR 65/05 GRUR 2010, 407 para. 16 ff.—One-piece eyelet; BGH 20.12.2011—X ZB 6/19, GRUR 2012, 378 para. 15 ff.—Installation device II; BGH 15.5.2012—X ZR 98/09, GRUR 2012, 803 para. 46—Calcipotriol monohydrate; BGH 25.9.2012—X ZR 10/10 GRUR 2013, 160 para. 30 f.—Toggle clamp device; BGH 11.3.2014—X ZR 139/10, GRUR 2014, 647 para. 26 f.—Ink supply system.

[74] BGH GRUR 2010, 407 one-piece eyelet; BPatG 2.2.2016 4 Ni 29/14 (EP); BPatG 28.11.2017 6 Ni 32/16 (EP); Fitzner/Lutz/Bodewig para. 38.

The could-would test is intended in particular to exclude an ex post facto analysis.[75] According to the could-would test, an invention is not considered non-inventive merely because the skilled person could have arrived at it based on the prior art, but only if there was sufficient reason to do so. This is intended to prevent a retrospective analysis in which much appears non-inventive.[76]

The skilled person does not act out of mere curiosity.[77] A skilled person is characterized by a goal-oriented approach. The skilled person is therefore guided by the technical result that he expects from a technical teaching.

5.13 Examination Scheme of the German Patent Office

The German and European Patent Offices have developed different examination schemes to assess inventive step. Examiners at the patent offices are required to apply the respective examination scheme in order to ensure predictability. In particular, this is intended to guarantee legal certainty and to exclude an ex post analysis.

The German examination scheme was developed by the Federal Court of Justice and comprises the following steps (Köllner, Mitteilungen der deutschen Patentanwälte, 5/2023, 202–210):

Step 1. Determine the technical field
Step 2. Identify the prior art
Step 3. Identify the disadvantages of the prior art
Step 4. The objective problem is derived from the disadvantages.
Step 5. Identify a first document as a starting point.
Step 6. Identify one or more further documents with which the invention can be obtained.
Step 7. Was there a reason for the skilled person to combine these documents? What speaks in favor and what speaks against it?

[75] EPO T 47/91; Singer/Stauder/Luginbühl Art 56 EPC para. 56; on its inadmissibility BGH GRUR 2001, 232 hole punch.

[76] EPO T 2/83 OJ EPO 1984, 265 = GRUR Int 1984, 527 simethicone tablet; EPO T 90/84; EPO T 124/84 EPOR 1986, 297; EPO T 223/84 EPOR 1986, 67; EPO T 256/84; EPO T 265/84 EPOR 1987, 193; EPO T 7/86 OJ EPO 1988, 381 = GRUR Int 1989, 226 xanthine; EPO T 392/86; EPO T 219/87; EPO T 274/87 EPOR 1989, 207; EPO T 564/89; EPO T 274/87 EPOR 1989, 207; EPO T 61/90; EPO T 513/90 OJ EPO 1994, 154 = GRUR Int 1994, 618 f. foamed bodies; EPO T 597/92 OJ EPO 1996, 135 = GRUR Int 1996, 814 rearrangement reaction; EPO T 167/93 OJ EPO 1997, 229 = GRUR Int 1997, 742 bleaching agent; EPO T 203/93; EPO T 406/98; BPatG 16.7.1997 20 W (pat) 64/95; Schulte para. 58 ff.; Szabo Mitt 1994, 225, 233 f.; Knesch VPP-Rdbr 1994, 70, 72.

[77] EPO T 0939/92 OJ 96, 309.

In this examination scheme, the invention under review is not used to work through the steps. The German examination scheme attempts to arrive at the invention independently. An ex post analysis, in which everything appears obvious in hindsight, is excluded. In the German examination scheme, it must be argued at each step what reason the skilled person has. In particular, it must be justified which document the skilled person would use as a starting point in step 5 and with which documents he would combine it in step 6.

5.14 Problem–Solution Approach of the European Patent Office

The examination scheme of the European Patent Office is called the problem–solution approach. The problem–solution approach serves to prevent an ex post analysis, in which everything appears very simple in hindsight.[78]

The problem–solution approach of the European Patent Office comprises the following steps:

Step 1: Identifying the closest prior art
Step 2: Determining the distinguishing features compared to the closest prior art.
Step 3: Identifying the technical effects to be achieved, determining the "objective technical problem" to be solved in order to arrive at the invention.
Step 4: Combination with additional prior art, if the could–would condition is fulfilled.

Regarding Step 1 The closest prior art is defined as the most promising starting point for achieving the invention under examination. The closest prior art should be in the same technical field, have a similar purpose or produce a similar effect, and require the fewest structural and functional modifications or share as many features as possible with the invention in order to arrive at the technical teaching of the invention. The closest prior art can therefore only be determined in light of the invention under examination, which corresponds to an ex post perspective.

The fact that the selection of the closest prior art actually constitutes an ex post application can already be seen from the fact that, if an invention appears obvious in view of an assumed prior art, the choice of the closest prior art cannot be questioned, since it has successfully led to the denial of inventive step.[79] The European Patent Office does not

[78] EPO T 24781; T 939/92; Szabo, Mitt. 1994, 226; Knesch, Mitteilungen der deutschen Patentanwälte 2000, 313; Schickedanz GRUR, 2001, 460 ff.; Stellmach, Mitteilungen der deutschen Patentanwälte 2007, 542 ff.

[79] EPO T 174/12; EPO T 824/05.

see this as an ex post issue. On the contrary, the European Patent Office considers the use of the prior art that comes closest to the invention to be an objective approach.[80]

The European Patent Office justifies the determination of the closest prior art in light of the invention by stating that this excludes arbitrariness in the selection of the starting point for the assessment of inventive step.[81] No argumentation is required as to why a document is assumed to be the closest, since it objectively shares the most features with the technical teaching of the invention.

Regarding Step 3 The objective technical problem is determined by identifying the features by which the closest prior art differs from the invention. The effect of the distinguishing features is established, and the objective technical problem is defined as the production of the effect resulting from the distinguishing features. Thus, in the third step as well, the invention under examination is used to work through a step of the examination scheme. This again results in an ex post issue, which is actually intended to be avoided.

Summary

In the problem–solution approach of the European Patent Office, the starting point is fundamentally the closest prior art.[82] In the selection of the closest prior art, the skilled person and whether they had any reason to regard this particular prior art as the closest play no role. The European Patent Office does not attempt to find a reason why this document in particular would be chosen by the skilled person as a starting point.

The skilled person on the filing or priority date is not the decisive factor in the selection of the closest prior art, but rather whether the chosen closest prior art can successfully demonstrate the obviousness of the invention under examination. In examination proceedings before the EPO, the applicant therefore does not have the opportunity to prove inventive step for their patent application by cleverly defining the skilled person.

It is not mandatory to apply the problem–solution approach. In particular, there is no legal basis that prescribes the application of the problem–solution approach for the

[80] EPO 6.4.1981—T1/80, OJ EPO 1981, 206; EPO 28.2.1984—T 181/82, OJ EPO 1984, 401; EPO 21.1.1986—T 248/85, OJ EPO 1986, 261; EPO 17.7.1986—T 164/83, OJ EPO 1987, 149; EPO 5.11.1987—T 254/86, OJ EPO 1989, 115, 123; EPO 18.9.1990—T 606/89, OJ EPO 1991, 24; EPO 24.9.1991—T 641/89, OJ EPO 1992, 25; EPO 25.11.1998—T284/94, OJ EPO 1999, 464.

[81] EPO 6.4.1981—T 1/80, OJ EPO 1981, 206; EPO 28.2.1984—T 181/82, OJ EPO 1984, 401; EPO 21.1.1986—T 248/85, OJ EPO 1986, 261; EPO 17.7.1986. T 164/83, OJ EPO 1987, 149, 154 f.; EPO 5.11.1987—T 254/86, OJ EPO 1989, 115, 123; EPO 18.9.1990—T 606/89, OJ EPO 1991, 24; EPO 24.9.1991—T 641/89, OJ EPO 1992, 25; EPO 25.11.1998—T 284/94, OJ EPO 1999, 464.

[82] EPO T BeckRS 2006, 30587752.

assessment of inventive step. However, non-application should be justified, since it is in fact the overwhelmingly predominant standard method in examination proceedings before the European Patent Office.

5.15 Differences between DPMA and EPO

The examination schemes of the DPMA and the EPO are similar. However, in examination proceedings before the DPMA, it is more common to argue from the perspective of the skilled person and to assert that the skilled person would not have used a particular document as the closest prior art. Figure 5.1 shows the EPO's examination scheme, which starts from the closest prior art in order to arrive at the invention. The closest prior art is defined independently of a skilled person as the document that has the most in common with the invention.

Figure 5.2 shows the examination scheme of the German Patent Office, which requires that the closest prior art be determined from the perspective of the skilled person (and not in comparison to the invention under examination).

Essentially, the difference between the German and European examination schemes is that, in the German procedure, it must be justified why the skilled person would consider a document to be the closest prior art. A justification is required for accepting a

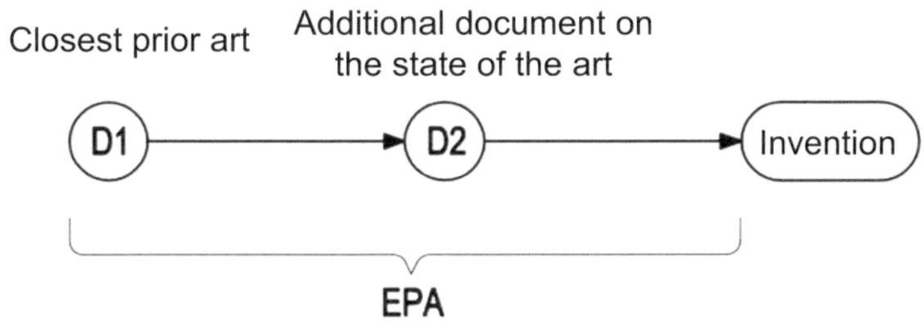

Fig. 5.1 EPO examination scheme

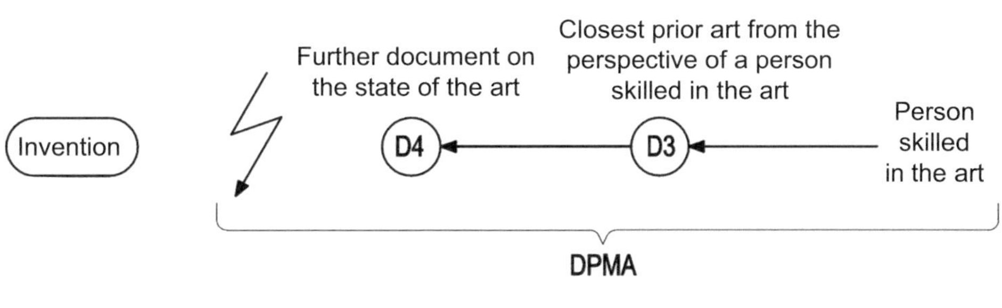

Fig. 5.2 DPMA examination scheme

document as the closest prior art.[83] In contrast, in the European procedure, the document that comes closest to the invention is determined as the closest prior art.[84] If no justification can be found that a document would be used by the skilled person as the closest prior art, this document cannot be used as the starting point for assessing inventive step in the German procedure. This provides an opportunity to argue for inventive step. This possibility does not exist in the European procedure.

The advantage of the European procedure is that, through the application of the strict problem-solution approach, the results of the European Patent Office are predictable. In contrast, the Federal Court of Justice, as the highest authority for examination proceedings before the German Patent Office, requires that each examination procedure be considered on a case-by-case basis. The expected outcomes in DPMA grant proceedings and contentious proceedings before the Federal Patent Court and the Federal Court of Justice are therefore only predictable with great uncertainty.

[83] BGH GRUR 2009, 382—Olanzapine; BGH GRUR 2009, 1039—Fish Bite Indicator; BGH GRUR 2017, 498—Knitted Shoe Upper.

[84] EPO T BeckRS 2006, 30587752; BeckOK PatR/Einsele, 33rd ed. 15.7.2024, EPC Art. 56 para. 5.

Indicators of Inventive Step

6

Contents

T. H. Meitinger, *Justification of the Inventive Activity*,
https://doi.org/10.1007/978-3-662-72735-5_6

Indicators, auxiliary criteria, auxiliary considerations, or circumstantial evidence can be used to substantiate the obviousness or non-obviousness of inventions. However, circumstantial evidence cannot replace the evaluative judgment of an examiner at the patent office or a judge involved in the case. Rather, they serve as decision-making aids.[1]

The assessment of whether an invention is inventive is an evaluative decision. All circumstances of the individual case must be taken into account. Circumstantial evidence is important in the decision and must be considered.[2]

[1] BeckOK PatR/Einsele, 33rd Ed. 15.7.2024, EPC Article 56 para. 9; BGH GRUR 79, 619—Tabelliermappe; BGH GRUR 62, 350—Dreispiegel-Rückstrahler; BGH GRUR 100,44—Dreinahtschlauchfolienbeutel; EPO T 0270/84; EPO T 1072/92; EPO T 0351/93.

[2] BGH GRUR 91, 129, Elastic Bandage.

If suitable circumstantial evidence is available, it should be asserted. However, these indicators are merely auxiliary considerations and can by no means be regarded as proof of inventive step. Rather, they can support the reasoning for inventive step or, conversely, help to invalidate a patent of an opposing third party. The application of the problem-solution approach or another examination scheme for assessing inventive step can generally not be dispensed with, even if circumstantial evidence is present.

6.1 Modification of Already Known Subject Matter

If something already known is adapted for a particular technical situation by routine activities, inventive step is generally not to be assumed.

6.2 Departure from Previously Conventional Practice

If the invention pursues a technical path not previously taken, even though a generally accepted path already existed, this is a strong indication of non-obviousness.[3]

6.3 Dimensions

The dimensions of a device can generally not establish inventive step.[4] However, if changing the dimensions opens up new technological fields or results in unexpected advantages, a sufficient inventive contribution may be present.[5]

6.4 Abstraction

If the invention consists in developing an abstracted form starting from a specific embodiment, the abstracted embodiment is obvious or not novel if the specific embodiment falls within the scope of protection ("The specific meets the general"). In the other case, non-obviousness must be present.

---

[3] BGH GRUR 99, 145—Shock Wave Lithotripter; BGH GRUR 09, 746—Operation of a Safety Device; EPO T 0221/85 OJ 87, 237.

[4] BGH GRUR 10, 814, Joint Sheets; BPatG Mitt. 84, 75.

[5] Huebner GRUR 2007, 839.

6.5 Foreseeable Difficulties

If the skilled person could expect difficulties with a particular solution, this speaks in favor of the inventive step of the invention that resulted from it.

6.6 Common General Knowledge

If a prior art can be established as common general knowledge, obviousness can be assumed. However, it must be proven that it actually constitutes the common general knowledge of the skilled person. Proof is successful if the common general knowledge can be found, in particular, in a textbook.

If an invention already results solely from the common general knowledge of the skilled person, the invention is obvious.

6.7 Analogous Application

An analogous application of known technical teachings is not inventive. However, if surprising and advantageous results can be achieved by the analogous application, inventive step is present. The evidential value of this indicator increases the longer the technical community has failed to recognize the advantages of the analogous application.[6]

6.8 Different Technical Field

If the inventive idea was found by transferring a technical teaching from a different technical field, this is an indicator of inventive step. However, if the other technical field is a neighboring technical field, this does not apply.

6.9 Number of Required Prior Art Documents

If a large number of prior art documents, at least four, are required to arrive at the invention, this speaks in favor of the inventive step of the invention. However, this does not apply if individual partial problems are solved in each case or if it was obvious for the skilled person to combine these documents.[7]

[6] BGH GRUR 69, 265 Disiloxan; BGH GRUR 66, 312 Appetite Suppressant I.

[7] EPO T 0315/88 OJ 90 Special Edition 27.

6.10 Problem Invention

In a problem invention, the invention results as the complementary element to the problem statement. In principle, the problem should be formulated in such a way that it contains no inventive elements. Therefore, no invention or parts thereof are to be seen in the problem statement.[8]

6.11 Adoption of the Invention by the Technical Community

If, shortly after the filing of the invention for patent, the invention is already discussed or further developed in additional applications, this speaks in favor of inventive step.[9]

6.12 Selection Invention

If, at the time of the invention, the skilled person had several alternatives to choose from, the selection of one alternative does not constitute an inventive achievement within the meaning of patent law. However, if a fortunate choice is made that leads to unexpected advantages, this rather speaks in favor of a sufficient inventive step.

If the invention can be achieved by a manageable number of experiments, there is no inventive step. However, if there is an unmanageable number of possibilities, the technical teaching may be inventive.[10] The selection of any embodiment that is not distinguished by special properties can in no case lead to inventive step.[11]

6.13 Automation/Computerization/Digitalization

Automation, computerization, or digitalization represent a general trend and can by no means serve as an indicator of inventive step.[12]

[8] BGH GRUR 84, 194 Rotary Harrow; BGH GRUR 85, 31 Acrylic Fibers.

[9] BGH GRUR 65, 473, 477 Permanent Waves.

[10] Mitt. 1937, 174, 175.

[11] BGH GRUR 2004, 47, 50 Bubble-Free Rubber Sheet I.

[12] EPO T 0775/90 OJ 93, Special Edition 28.

6.14 Known Bonus Effect

If an advantageous effect is known and this effect is used for an existing device, this does not lead to inventive step, but is attributable to the skills of the average skilled person.[13]

6.15 Urgent Need

Satisfying an urgent need is a strong indication of inventive activity. This is especially true if the need has existed for a long time and no suitable solution has been found.[14]

If the invention addresses a long-standing need that the technical community has so far been unable to adequately satisfy, this is a strong indication of inventive activity.[15] The longer the period during which the need could not be met, or not adequately met, the more convincing the evidence becomes.[16]

On the other hand, if a need has only recently arisen, this may prompt the skilled person to examine devices and methods from other documents for their applicability. In such cases, it is more likely that combining documents from the prior art would be considered obvious.

6.16 Simplicity of the Invention

If the invention provides a simple technical solution to a problem that previously could only be solved by complicated technology, this is an indication of non-obviousness. The skilled person is generally motivated to achieve solutions that are easy to manufacture and therefore cost-effective. The fact that this was achieved by the invention, while the general technical community in the field failed to do so, is strong evidence of inventive activity. This is particularly true if it was only achieved after a considerable period of time.[17]

---

[13] BGH GRUR 03, 317 Cosmetic Sunscreen; BGH GRUR 03, 693 High-Pressure Cleaner; BGH GRUR 09, 936 Heater.

[14] BGH BIPMZ 53, 227—Pipe Clamp; BGH BIPMZ 58, 114—Upholstered Chair; BGH GRUR 62, 350—Three-Mirror Reflector; BGH GRUR 70, 289—Slide Mount IV; BGH GRUR 79, 619—Tabulating Folder.

[15] BGH BIPMZ 1953, 227—Pipe Clamp, BGH BIPMZ 1973, 257—Herbicide; BGH GRUR 70, 289, 294 Slide Mount IV.

[16] BGH GRUR 1969, 182, Concrete Dispenser; BGH GRUR Circuit Arrangement.

[17] BGH BIPMZ 79, 151—Labeling Device II; BGH GRUR 99, 145—Shock Wave Lithotripter, EPO T 0106/84 OJ 85, 132; EPO T 0009/86 OJ 88,12; EPO T 0229/85 OJ 87, 237; Mitt 78, 136 Oil Pipe.

6.17 Expectation of Success

If the development of an invention is associated with an expectation of success, such that the inventor could assume that the creation of the invention would succeed, this generally does not indicate inventive achievement.[18]

A sufficient expectation of success can lead to the obviousness of an invention. The expectation of success depends on the technical field, the effort required to pursue the approach, the availability of alternatives, and the prospects for economic return.[19]

If the skilled person could reasonably expect success with a particular approach, this generally indicates that the inventive step required for patentability is lacking.

An expectation of success exists even in the presence of typical problems when transferring a method or device from another technical field. Common, foreseeable problems cannot preclude obviousness.

6.18 Non-Analogous Prior Art

If only non-analogous prior art exists, and especially if the closest prior art is non-analogous, this suggests inventive activity.

6.19 Serendipity/Chance

The subjective effort or labor involved in arriving at an invention must be disregarded when assessing inventive activity. A serendipitous discovery does not preclude inventive activity.[20] However, inventive activity is not present if the inventive teaching was found through straightforward experiments from a limited number of possibilities.[21]

If an invention is achieved by chance, this is not a negative indication.[22] The manner in which an invention was created is irrelevant. A patent does not reward the inventor's effort, but rather the enrichment of technical knowledge. Even the recognition that an invention exists can constitute inventive activity.

[18] BGH GRUR 01, 730 Trigonelline; EPO T 0060/89 OJ 92, 268; EPO T 0386/94 OJ 96, 658; EPO T 0412/93 EPOR 95, 629.

[19] BGH GRUR 2012, 803—Calcipotriol Monohydrate; BGH GRUR 2016, 1027—Celiac Disease Diagnostic Method; BGH GRUR 2019, 1032—Fulvestrant; BGH GRUR 2020, 521—Autoantibody Detection; BGH GRUR 2020, 602—Tadalafil; BGH GRUR 2020, 1178—Pemetrexed II; BGH GRUR 2021, 696—Phytase; BGH GRUR 2024, 236—Sorafenib Tosylate.

[20] BGH GRUR 96, 757 Gear Cutter.

[21] BGH GRUR 92, 375, Tablet Disintegrant.

[22] EPO T 0356/93 OJ 95, 545.

6.20 Technical Skill/Constructive Measures

Technical skill is part of the expertise of a skilled person. If an invention is developed solely from established rules, the technical teaching of the invention is considered obvious.[23]

Purely technical measures never constitute a patent-relevant inventive activity.[24] If the technical or constructive skill of the average skilled person is not exceeded, the relevant technical teaching cannot be patented. This also applies if the technical teaching is practical, clever, and advantageous for manufacturing.[25]

Purely constructive measures fall within the expertise of the average skilled person and cannot establish inventive activity. However, if the invention is based on several constructive considerations, an invention may have been created.[26]

6.21 High Level of Development Activity

An invention belonging to a technical field with a high level of general development activity is more likely to be considered inventive.[27]

6.22 Young Technical Field

For an invention in a young technical field, it must be considered that there may be too few documents available to adequately assess the invention.[28]

6.23 Commercial Achievement

All efforts and results that are not of a technical nature cannot establish inventive activity. If only market needs are identified and met using known technologies, there is no invention within the meaning of the Patent Act.[29]

---

[23] BGH Three-Mirror Reflector; BGH GRUR 62, 80 Pipe Seal.

[24] BGH GRUR 1954, 107, 110—Multiple Clamp; BGH GRUR 1954, 258, 259 Channel Iron; BGH GRUR 1956, 73, 76—California Shoes.

[25] BGH 1971/1973—Breathing Mask.

[26] BGH Mitt 72, 18 Electric Shaver; BGH GRUR 87, 351 Wall Box II; BGH BIPMZ 79, 151 Labeling Device II.

[27] EPO T 0009/86 OJ 88, 12; EPO T 0229/85 OJ 87, 237, No. 7.

[28] EPO T 0500/91 EPOR 95, 69.

[29] BGH GRUR 90, 594 Computer Carrier.

6.24 No Prior Art

If no or only minimal prior art can be identified, a groundbreaking invention may be present, which regularly fulfills the criterion of inventive activity. An unsuccessful search for prior art should in any case be regarded as an indication of patentability.

6.25 Kinematic Reversal

A kinematic reversal, or kinematic inversion, occurs, for example, when in a device, movable parts are made stationary and previously immovable parts are made movable, or when the directions of movement of components are swapped. In kinematic reversal, the same effect is generally intended. Such a kinematic reversal falls within the professional competence of the skilled person and does not constitute an inventive step.

6.26 Combination Invention

A mere juxtaposition of features, each of which is known, does not result in an inventive step. An exception exists if the combination is particularly difficult or especially advantageous and leads to a significant benefit.[30] If, moreover, the skilled person had no reason to combine precisely these features, an inventive step is also more likely to be assumed.[31]

A combination invention arises if features already known from the prior art are combined in such a way that synergy effects result. A combination invention is inventive if the skilled person has not received any suggestion from the prior art to combine these features.[32]

6.27 Cost-Effective Production

A reduction in manufacturing costs, especially for mass-produced items where cost-effective production is particularly significant, is an indication of non-obviousness.

[30] BGH BIPMZ 63, 365, 366 Schutzkontaktstecker.

[31] BGH GRUR 69, 182 Betondosierer; BGH GRUR 81, 736 Kautschukrohlinge; BGH GRUR 99, 145 Stoßwellen-Lithotripter.

[32] BGH GRUR 69, 182 Betondosierer; BGH GRUR 81, 736 Kautschukrohlinge; BGH GRUR 99, 145 Stoßwellen-Lithotripter.

6.28 Cumulation of Features

The mere addition of unrelated features does not make an aggregation inventive. However, if the combination of features leads to a particular advantage and special technical hurdles had to be overcome, an inventive step may be present.[33]

6.29 Long Time Period

If a particularly long period has elapsed between the emergence of the technical problem and its solution, and there have been extensive efforts by the technical community to find a solution, an inventive step may be present.[34]

If, in particular, there was an unsatisfied need over a long period that was ultimately solved by the invention, an inventive step is present.[35] This applies all the more if the technical community has been striving for a suitable solution for some time.[36]

6.30 Granting of Licenses

The granting of licenses is an indication of the patentability of an invention, since a market participant will only pay license fees for patents with inventions that they consider to be legally valid.[37]

6.31 Praise from the Technical Community

A positive evaluation by experts can be considered an indication of inventive step.[38]

[33] BGH BIPMZ 63, 365, 366 Schutzkontaktstecker; BGH BIPMZ 60, 87, 91 electromagnetic stirring device.

[34] BGH BIPMZ 54, 24—Mehrfachschelle; BGH GRUR 57, 488—Schleudergardine; BGH GRUR 62, 290—Brieftaubenreisekabine; BGH BIPMZ 53, 227—Rohrschelle; BGH BIPMZ 58, 114—Polstersessel; BGH GRUR 65, 416—Schweißelektrode; BGH BIPMZ 89, 133—Gurtumlenkung; BGH GRUR 96, 757—Zahnkranzfräser; EPA T 0109/82 ABl. 84, 473; EPA T 0090/89 GRUR-In 91, 81.

[35] BGH GRUR 57, 488 Schleudergardine; BGH GRUR 62, 290 Brieftaubenreisekabine.

[36] BGH GRUR 60, 427 Fensterbeschläge; BGH GRUR 59, 22 Einkochdose.

[37] BGH decision of 12.12.2000—X ZR 121/97 Knee Joint Endoprosthesis.

[38] EPA T 0106/84 OJ 85, 132; EPA T 0677/91 OJ 94 Special Edition 43.

6.32 Mass-Produced Articles

A large sales success is evidence of a significant market need that was ultimately satisfied. However, if the market success is not based on the technical teaching realized by the product, but rather on a clever marketing campaign, the market success is not evidence of inventive step.

For mass-produced articles, there is increased development pressure, so that even small improvements may be based on an inventive step. These improvements must lead to technical advantages that justify the commercial success.[39]

6.33 Choice of Material

The skilled person is familiar with the properties of various materials, so a mere choice of material does not constitute an inventive step. There is also no inventive achievement if the relevant average skilled person does not know the properties of a material but learns them from an expert.[40]

6.34 Multiple Steps

If several steps are required to move from the prior art to the invention, and these steps are not merely routine for the skilled person, a sufficient inventive step for patent grant may be present.[41]

6.35 Multiple Independent Inventions

If an invention is created multiple times and independently, this is an indication that the invention was obvious.[42]

---

[39] BGH BIPMZ 55, 66 Polsterkörper Latex; BGH GRUR 82, 406, 409 Verteilergehäuse.

[40] BGH GRUR 62, 350 Dreispiegel-Rückstrahler.

[41] BGH GRUR 78, 98, 99 Schaltungsanordnung; BGH GRUR 81, 190, 193 Skistiefelauskleidung; BGH GRUR 85, 369 Körperstativ; BGH Mitt 04, 69 Ankerwickelmaschine; BGH decision of 10.7.2007—X ZR 240/02 Folding Box.

[42] BGH GRUR 53, 384 Zwischenstecker I; BGH GRUR 81, 341 piezoelectric lighter; BGH GRUR 53, 120 Glimmschalter.

6.36 Rights of Joint Use

If competitors seek rights of joint use, this can be seen as an indication of inventive step, since a competitor will not make efforts if they believe the relevant patent is not legally valid.[43] However, if it requires little effort and cost for the competitor to obtain a right of joint use, this is at most a weak indication. A right of joint use is, in particular, a simple, non-exclusive license.

6.37 Imitation

If a new technical teaching is used by competitors, this speaks in favor of the patentability of the invention in question.[44] However, this does not apply if competitors use the technical teaching not for technical reasons, but for marketing reasons.[45]

If the invention is imitated by competitors and previously used technologies are replaced by the invention, this is strong evidence of an inventive achievement.[46]

6.38 Subsequent Inventions

If subsequent inventions are based on the invention under evaluation, this is evidence of inventive step.[47]

6.39 Disadvantages of the Prior Art

The mere existence of individual disadvantages inherent in a technical teaching does not render the use of the technical teaching of the prior art inventive for the realization of an invention.[48]

[43] EPA T 0351/93 OJ EPO 96 Special Edition 32.

[44] BGH GRUR 91, 120 Elastic Bandage; BGH GRUR 87, 351 Mauerkasten II; BGH GRUR decision of 28.1.1997 X ZR 43/94 Rearview Mirror.

[45] BGH GRUR 91, 120 Elastic Bandage.

[46] BGH GRUR 91, 120 Elastic Bandage; BGH GRUR 87, 351 Mauerkasten II; BGH of 28.1.97 X ZR 43/94 Rearview Mirror.

[47] BGH GRUR 65, 473, 477 Dauerwellen.

[48] BGH judgment of 15 June 2021—X ZR 58/19, GRUR 2021, 1277—Führungsschienenanordnung; BGH judgment of 3 September 2024—X ZR 106/22—Scheibenbremse III.

6.40 New Approach

If the invention has taken a new approach, this indicates inventive step. The new approach should have led to an alternative solution or a better technical teaching in order to establish inventive step.[49]

6.41 Utility

Particularly high utility can be a positive indication.

6.42 Obvious-to-try

A person skilled in the art is expected to conduct experiments if there is a reasonable prospect of success. This does not apply if a large number of experiments are required to create an invention, or if the direction of the experiments runs counter to a general trend.[50]

6.43 Optimization

Optimization is a routine task for a skilled person and cannot establish inventive step. However, if optimization is achieved through the targeted selection of a particular technical teaching and this results in a surprising outcome, this is evidence of inventive achievement.

Optimization of two conflicting parameters represents the usual expertise of a skilled person. This is especially true if the skilled person can calculate the optimal setting or determine it through simple experiments.[51]

6.44 Parallel Applications

If the same invention is filed in parallel in several applications by different applicants, and the technical teachings of the parallel applications represent a divergent and more cumbersome or disadvantageous solution, this is an indication of the patentability of the invention in question.[52]

[49] BGH GRUR 09, 746 Betrieb einer Sicherheitseinrichtung.

[50] T 0253/85 EPOR 87, 198, 204; T 0348/86 EPOR 88, 159.

[51] BGH GRUR 13, 1022 Aufzugsmultigruppensteuerung.

[52] BGH GRUR 60, 427 Fensterbeschläge; BGH GRUR 53, 384 Zwischenstecker I; BGH GRUR 53, 120 Glimmschalter.

6.45 Planned and Systematic Work

How the invention was created is fundamentally irrelevant. Whether a systematic and planned approach or chance led to the invention plays no role in the assessment of inventive step. Only the objective contribution to the advancement of technology is to be evaluated, and this contribution must go beyond an evolutionary development of the technology.

6.46 Practical Proven Effectiveness

If an invention has already proven itself well in practice, this is an indication of inventive step.[53]

6.47 Routine

Technical difficulties that can be solved by routine work cannot lead to inventive step within the meaning of the Patent Act.[54]

6.48 Skilled Person's Toolkit

Assertions such as "it was obvious" or that it belonged to the "everyday toolkit of the skilled person" must be substantiated in order to assume a lack of inventive step.

6.49 Material Substitution

If one material is replaced by another and the skilled person is aware of the other material and its advantageous properties, there is no inventive step.[55] This does not apply if a new property of the substitute material is discovered through the invention.

[53] BGH GRUR 59, 22, Einkochdose.

[54] T 0073/88 ABl 91 SonderA 23.

[55] BGH GRUR 62, 80 Rohrdichtung; BGH GRUR 67, 25 Spritzgussmaschine III; BGH GRUR 62, 350 Dreispiegel-Rückstrahler; BGH GRUR 10, 814 Fugenglätter; BGH GRUR Int 10, 334 Sektionaltor.

6.50 Technical Progress

Technical progress is strong evidence of inventive step, because if the invention had been obvious, any skilled person would have used the invention to achieve technical progress.[56]

Technical progress is achieved when a new means is provided that produces a useful result. Therefore, an alternative solution to an already existing invention does not constitute technical progress. However, if the alternative solution offers advantages over the previously known technical teaching, then the alternative solution also constitutes technical progress.[57]

If the invention in particular achieves significant technical progress, this is very strong evidence of inventive step.[58]

6.51 Technical Prejudice

If the subject matter of the patent application represents a departure from what has previously been customary, this can be regarded as strong evidence of inventive step.[59] If relevant textbooks or other publications can be identified that demonstrate the invention contradicts the prevailing teaching, it can hardly be considered obvious.

A technical prejudice is a general opinion held by the majority of experts, which prevents research and development in the direction of the invention.[60] A technical prejudice exists if:

- the prejudice is generally held in the technical community[61],
- the technical community therefore considers the invention not to be feasible[62],
- the prejudice proves to be an error as a result of the invention[63] and
- the prejudice is technical and not economic in nature.[64]

If the invention additionally produces a technical effect or technical advantage, inventive step may be present.

---

[56] BGH Gießpulver; BPatG Mitt 87, 10; EPA T 0164/83 ABl 87, 149; EPA T 0181/82 ABl 84, 401, No. 4; EPA T 0095/83 ABl 85, 75.

[57] BGH Anthradipyrazol.

[58] BGH BIPMZ 89,215—Gießpulver; BGH GRUR 94, 36—Messventil.

[59] BGH GRUR 99, 145, Stoßwellen-Lithotripter.

[60] BGH GRUR 96, 857 Rauchgasklappe; BGH GRUR 84, 580, 581 Chlortoluron.

[61] BGH GRUR 96, 857 Rauchgasklappe.

[62] BGH Liedl 61/62, 397, 411 Straßenbeleuchtung; BGH GRUR 96, 857 Rauchgasklappe.

[63] BGH Liedl 61/62, 397, 411 Straßenbeleuchtung; BGH GRUR 96, 857 Rauchgasklappe.

[64] BGH GRUR 94, 36 Messventil.

A relevant technical prejudice exists if it was present in the technical community at the time of the filing or priority date of the patent application.[65]

6.52 Trial and Error/Experiments

If the skilled person arrives at the invention through manageable experiments, this generally does not constitute inventive activity. However, this only applies if the skilled person had a reason to conduct the appropriate experiments. If experiments are used merely to determine the best solution among several technical teachings, this falls within the routine skills of the skilled person and is considered obvious.[66]

6.53 Transfer Invention

The mere transfer of a known theory to a similar problem cannot establish inventive activity. However, if the professional community advises against such a specific transfer, or if the transfer is made to a completely different technical field, a different assessment may result.

6.54 Use

An invention can realize a novel use. If the type of use is not obvious, inventive activity is present.[67] If a previously unknown method of implementation is required for the novel use, inventive activity can be assumed with certainty.

6.55 Advantages

If an invention produces advantages, this can be evidence of inventive activity, since every company strives to generate advantages for itself. The fact that other companies did not create the invention can be considered proof of non-obviousness. The advantage can also be minor, as even small advantages can have major economic consequences.[68]

[65] BGH Liedl 61/62, 397, 411 Straßenbeleuchtung; BGH GRUR 96, 857 Rauchgasklappe.

[66] BGH GRUR 68, 311, 313 Garmachverfahren; BPatG Mitt 65, 10.

[67] BGH GRUR 12, 373 Glass fibers.

[68] EPA T 0015/86 EPOR 87, 291.

6.56 Arbitrary Selection from a Multitude of Variants

An arbitrary selection of an embodiment from a multitude of possible embodiments, which does not stand out from the alternative embodiments due to particularly advantageous properties, cannot establish inventive activity.

6.57 Economic Success

If economic success is attributable to the technical teaching and not to business management efforts, this may indicate inventive activity.[69]

If licenses are granted for the use of an intellectual property right, this is a positive indication, since license fees are generally only paid for legally valid intellectual property rights.[70]

If economic success over a longer period can be regarded as the market's judgment of an invention, this is strong evidence of inventive achievement. However, if the economic success is based solely on a marketing concept, the economic success cannot be given any significance.[71]

6.58 Time as a Criterion

If only very old documents can be found relating to an invention, this is an indication of inventive activity, since it was apparently difficult to arrive at the technical teaching of the invention.[72] However, this only applies if there was a need during this period that is satisfied by the invention.[73]

A document that is already 20 or 30 years old would not be consulted by a skilled person in a dynamic technical field. Therefore, it is not obvious for them to use this document to arrive at an inventive idea. An example of a dynamic technical field is telecommunications. On the other hand, a skilled person in a technical field not characterized by frequent technological upheavals will certainly also consider documents that are 40 or 50 years old. A conservative technical field, for example, is general mechanical engineering.

In particular, it is unlikely that the skilled person would use a document that is already more than 20 years old as the closest prior art. The skilled person will always strive to

---

[69] BGH GRUR 91, 120—Elastic Bandage; BGH GRUR 90, 594—Computer Carrier; BGH GRUR 87, 351—Wall Box II; BGH BIPMZ 94, 36—Measuring Valve.

[70] BGH of 12.12.2000 X ZR 121/97 Knee Joint Endoprosthesis.

[71] BGH GRUR 91, 120 Elastic Bandage; BGH GRUR 90, 594 Computer Carrier.

[72] BGH GRUR 57, 488 Centrifugal Curtain; BGH GRUR 62, 290 Carrier Pigeon Travel Cabin.

[73] BGH GRUR 65, 416 Welding Electrode I; BGH GRUR 96, 757 Gear Cutter.

use the current state of knowledge in their field to find a suitable solution to a technical problem. Therefore, a current document should be used as the closest prior art. However, if only a document older than 20 years can be used as the closest prior art, this speaks in favor of inventive step.

6.59 Inevitable Development

If a new technical teaching can be evaluated as a consistent further development of existing technology, especially since no alternatives can be identified, the technical teaching cannot be granted inventive activity.[74]

If an invention arises almost automatically as a result of technological advancement, with alternatives appearing unlikely, inventive step can be excluded.[75] Even if a particularly advantageous effect occurs in this case, this cannot establish inventive activity.

If the skilled person inevitably arrives at an invention through an obvious process, inventive achievement cannot be present.[76] In principle, the Patent Act is not intended to be accessible for such inventions that result almost automatically from the inevitable, consistent further development of technology.

6.60 Practical Measures

Practical measures that result merely from the application of known considerations and are not based on fundamental deliberations are attributed to the skills of the average skilled person and cannot establish inventive step. This is especially true if only minor changes or improvements are made. If, for reasons of practicality, already known ideas whose effects are predictable are applied, it also cannot be assumed that there is a sufficient inventive step for patent granting.

---

[74] BGH GRUR 09, 746 Operation of a Safety Device.

[75] EPA T002/83 OJ 84, 265, No. 6; EPA T 0192/82 OJ 84, 415, No. 16.

[76] BGH GRUR 12, 1130 Leflunomide.

Practical Examples

7

Contents

The assessment of inventive step, in contrast to the assessment of novelty, is not self-evident. For the justification of novelty, the following statement is generally sufficient: "The features XY of the main claim are not contained in the prior art." No further argumentation is required. The justification of inventive step is considerably more complex and requires appropriate and suitable argumentation. A targeted approach is necessary to identify the technical problem to be solved, as it must not be too specific, so that the invention does not appear as an inevitable consequence of the problem statement. Strategy is required if it becomes necessary to add features to the original main claim, since these distinguishing features form the starting point for the entire subsequent justification of inventive step.

T. H. Meitinger, *Justification of the Inventive Activity*,
https://doi.org/10.1007/978-3-662-72735-5_7

7.1 Basic Approach

Figure 7.1 illustrates the basic approach to arguing the presence of inventive step. The argumentation for a lack of inventive step must be adapted accordingly. If, from the perspective of the skilled person, the relevant prior art documents can be combined, the (amended) main claim must contain one or more features that are not described in the prior art documents. These distinguishing features serve to determine the effect or function of the invention. The function of the distinguishing features should be formulated in such a way that it cannot be derived from the prior art. The function of the invention gives rise to the problem to be solved, which should be formulated in general terms. By formulating the problem in general terms, it is ensured that the invention does not inevitably result from the problem itself. If the prior art documents can be combined, it is therefore necessary to create sufficient distance from the prior art by means of

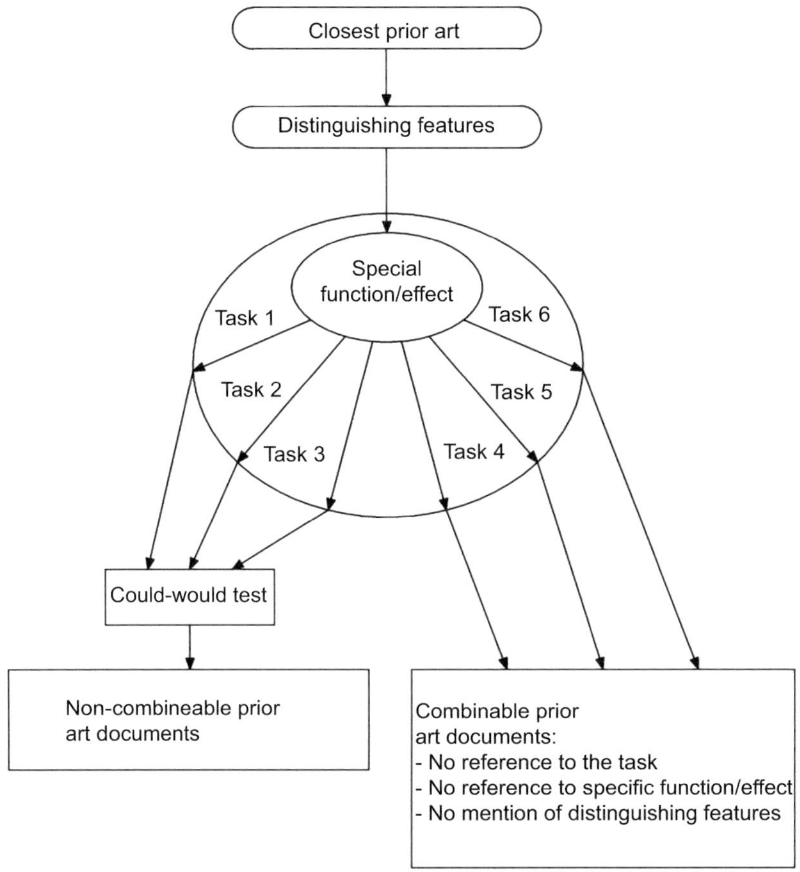

Fig. 7.1 Argumentation of inventive step

distinguishing features that fulfill a different function and problem. If the prior art documents cannot be combined, this must be demonstrated using the could-would test, and the argumentation could end here. However, as a precaution, the distance to the prior art could be further increased by incorporating additional features into the main claim. At the very least, a description of the specific function and problem to be solved by the distinguishing features can provide an additional argumentative distinction from the prior art. The justification of inventive step must be carried out in the same way for all independent claims.

If the distinguishing features can be derived from a combination of prior art documents, the could-would condition must be used for argumentation. In this case, it must be demonstrated that the skilled person would not combine these prior art documents.

The following general notes should be observed:

- Where possible and appropriate: Broaden the main claim with one or more features that are not found in any prior art document.
- If applicable: Broaden the main claim with one or more features that are only found in prior art documents that the skilled person would not combine with the closest prior art. Argumentation based on the could-would test should follow.
- Determine the specific function or effect of the distinguishing features in such a way that the function and effect are not disclosed in the prior art.
- Define the problem to be solved in such a way that the invention does not logically follow from the problem statement.
- It is advantageous if the problem to be solved is not described in the prior art.

7.2 Closest Prior Art

The closest prior art is usually specified by the examiner in the first office action of the patent office. In European proceedings, it makes no sense to attempt to propose a different document as the closest prior art by defining the skilled person in such a way as to replace the document that shares the most features with the invention under examination. The situation is more promising in German patent grant proceedings. In proceedings before the DPMA, the starting point is generally the average skilled person, so with suitable argumentation regarding the skilled person, a desired prior art document can be justified as the closest prior art. Of course, it is always questionable whether the examiner at the patent office will follow the argumentation.

7.3 Distinguishing Features

The distinguishing features are those features that are not found in the closest prior art. Ideally, these features cannot be derived from any other prior art document either.

7.4 Problem to be Solved by the Invention

In the course of the patent grant procedure, the main claim and the other independent claims are usually broadened by additional features to distinguish them from the prior art. This may also necessitate changing the problem to be solved. It is permissible to adapt the problem to be solved by the invention during the patent grant procedure.

7.5 Example 1: Device for Sieving Compost

Patent application DE 10 2022 114 831 "Device for Sieving Compost" was filed with the patent office on June 13, 2022, and resulted in a patent on July 27, 2023. The technical teaching of the patent describes a sieve that can be movably arranged, for example, on a wheelbarrow, with compost being filled into the sieve and shaken. Stones and large roots are retained by the sieve, and the humus can be collected in the loading area of the wheelbarrow.[1]

The problem described in the original patent application was as follows:

"The problem is therefore to provide a device by which the sieving of compost to produce humus is simplified. In particular, possible health hazards should be excluded."[2]

The problem was therefore originally argued as the simplification of humus production in order to avoid health risks due to heavy physical labor.

The original main claim read as follows:

"Device for sieving compost, comprising: a container (4) for receiving compost, wherein the bottom of the container (4) is at least partially designed as a sieve, and a rail (2), wherein the container (4) can be moved on the rail (2)."[3]

According to the invention, the sieve no longer had to be held by hand, but could be placed on rails, thereby relieving strain on the back and thus preventing health risks.

Figure 7.2 shows the device according to the invention.

The patent office was able to identify two prior art documents that come very close to the invention. The closest prior art is FR 2 692 820, which discloses all features of the original claim of the application. Figure 7.3 shows Fig. 1 of FR 2 692 820 with a sieve on rails, which rest on the tub of a wheelbarrow, and the sieve is moved back and forth by means of a lever.

[1] DPMA, https://depatisnet.dpma.de/DepatisNet/depatisnet?action=pdf&docid=DE102022114831B3&xxxfull=1, paragraphs [0005] and [0015], accessed on 4.10.2024.

[2] DPMA, https://depatisnet.dpma.de/DepatisNet/depatisnet?action=pdf&docid=DE102022114831B3&xxxfull=1, paragraph [0004], accessed on 4.10.2024.

[3] DPMA, https://depatisnet.dpma.de/DepatisNet/depatisnet?action=pdf&docid=DE102022114831B3&xxxfull=1, summary, accessed on 4.10.2024.

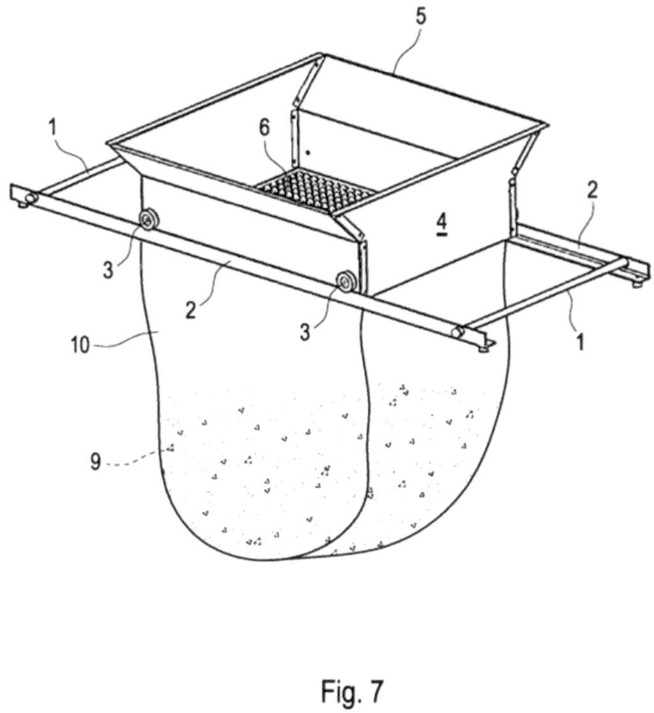

Fig. 7.2 Figure 7 of DE102022114831B[4]

In Fig. 7.3, all features of the main claim are disclosed, making it necessary to include an additional feature in the main claim to distinguish it from FR 2 692 820 A1.[6]

In addition, the patent office also researched US 2021/0362190 A1, which discloses a receptacle with a sieve as the bottom surface. Figure 7.4 shows Fig. 1 of US 2021/0362190 A1.[7]

The main claim of the patent application therefore had to be amended. The patentable main claim reads as follows:

1. *"Device for screening compost, comprising:*
   a container (4) for receiving compost, wherein the bottom of the container (4) is at least partially designed as a sieve and a rail (2), wherein the container (4) can be

---

[4] DPMA, https://depatisnet.dpma.de/DepatisNet/depatisnet?action=pdf&docid=DE102022114831B3&xxxfull=1, accessed on 29.5.2025.

[6] DPMA, https://depatisnet.dpma.de/DepatisNet/depatisnet?action=pdf&docid=FR000002692820A1&xxxfull=1, accessed on 2.1.2025.

[7] DPMA, https://depatisnet.dpma.de/DepatisNet/depatisnet?action=pdf&docid=US020210362190A1&xxxfull=1, accessed on 31.12.2024.

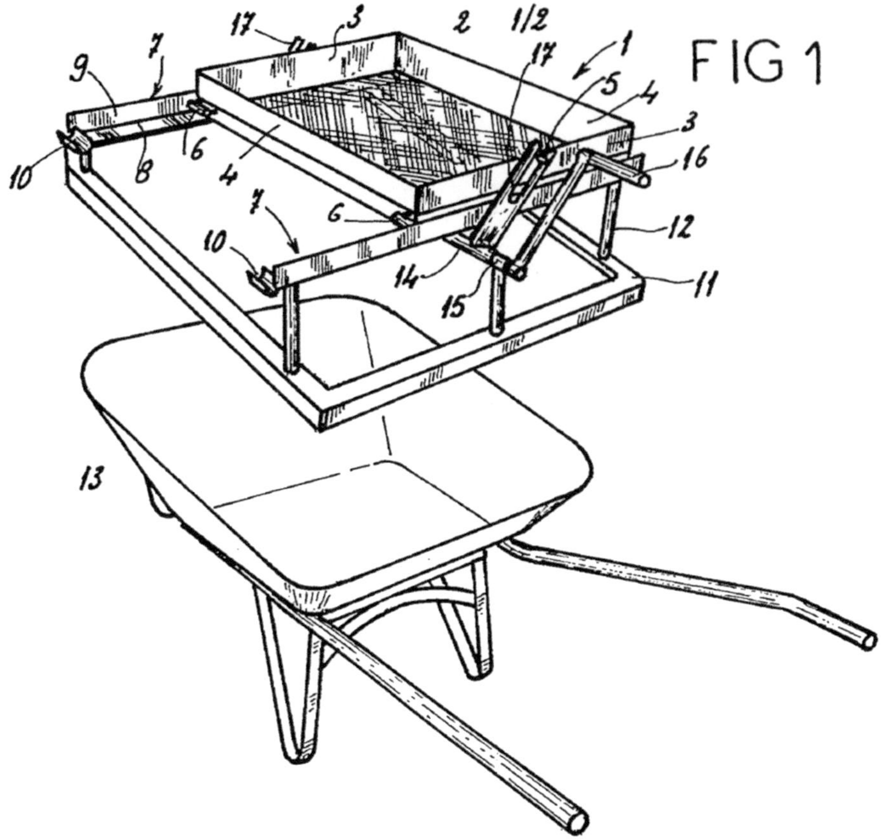

Fig. 7.3 Figure 1 of FR2692820A1[5]

moved on the rail (2), wherein ***any number of rollers (3) or runners are arranged on the container (4), on which the container (4) is movable on the rail(s) (2).***"[9]

The new features enable movement of the sieve with low friction. The objective of achieving a health-promoting effect or a back-friendly working method could no longer be pursued, since the prior art also disclosed placing the sieve on a wheelbarrow. The function of the additional features was to reduce friction during movement of the sieve. This function could not be directly derived as the objective, since a problem statement such as "reducing friction of the sieve during movement" would have suggested the use

[5] DPMA, https://depatisnet.dpma.de/DepatisNet/depatisnet?action=pdf&docid=FR000002692820A1&xxxfull=1, accessed on 29.5.2025.

[9] DPMA, https://depatisnet.dpma.de/DepatisNet/depatisnet?action=pdf&docid=DE102022114831B3&xxxfull=1, accessed on 31.12.2024.

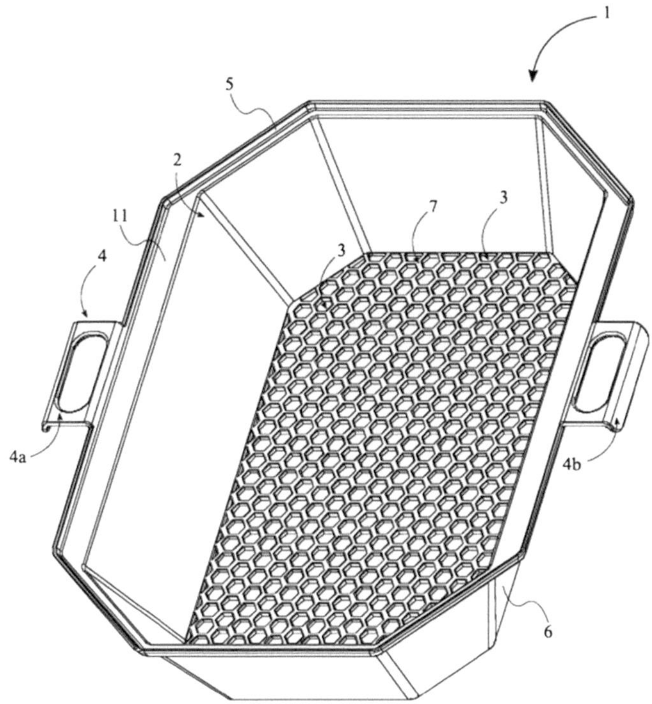

FIG. 1

Fig. 7.4 Figure 1 of US20210362190A1[8]

of rollers. In this case, the technical solution would have been attributed to the skilled person's routine abilities. The problem therefore had to be abstracted and described as "efficient and rapid screening." With this objective of "efficient and rapid screening," it was possible to obtain a patent grant.

7.6 Example 2: Riding Crop

The patent DE 10 2022 105 981 "Riding Crop" was filed with the patent office on March 15, 2022. Figure 7.5 shows the riding crop according to the invention.

[8] DPMA, https://depatisnet.dpma.de/DepatisNet/depatisnet?action=pdf&docid=US020210362190A1&xxxfull=1, accessed on 29.5.2025.

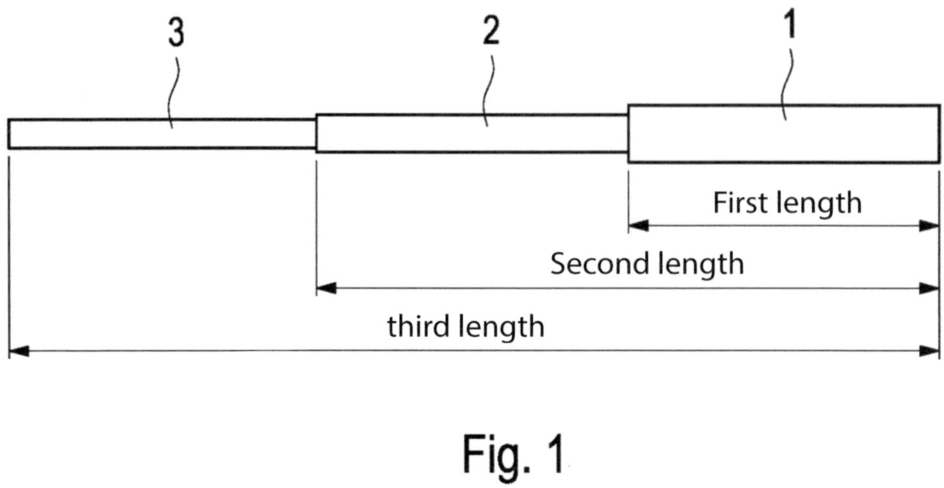

Fig. 7.5 Figure 1 of DE102022105981A1[10]

The main claim of the originally filed patent application read as follows:

1. *"Riding crop as equipment for a rider comprising: a handle as a first section (1) and a second section (2), characterized in that the length of the first section (1) together with that of the second section (2) corresponds to the prescribed length for a first discipline of an equestrian competition."*[11]

The German Patent Office was able to identify two highly relevant prior art documents, namely DE 20 2020 106 967 U1 and FR 2 279 438.

DE 20 2020 106 967 U1 describes a length-adjustable riding whip, the length of which can be changed.

Figure 7.6 shows the riding crop of DE 20 2020 106 967 U1. The length of the riding crop can be varied telescopically.[12]

Figure 7.7 shows the riding whip of the second prior art document, the length of which can also be adjusted telescopically.[14]

---

[10] DPMA, https://depatisnet.dpma.de/DepatisNet/depatisnet?action=pdf&docid=DE102022105981A1&xxxfull=1, accessed on 29.5.2025.

[11] DPMA, https://depatisnet.dpma.de/DepatisNet/depatisnet?action=pdf&docid=DE102022105981A1&xxxfull=1, accessed on 31.12.2024.

[12] DPMA, https://depatisnet.dpma.de/DepatisNet/depatisnet?action=pdf&docid=DE202020106967U1&xxxfull=1, accessed on 2.1.2025.

[14] DPMA, https://depatisnet.dpma.de/DepatisNet/depatisnet?action=pdf&docid=FR000002279438A1&xxxfull=1, accessed on 2.1.2025.

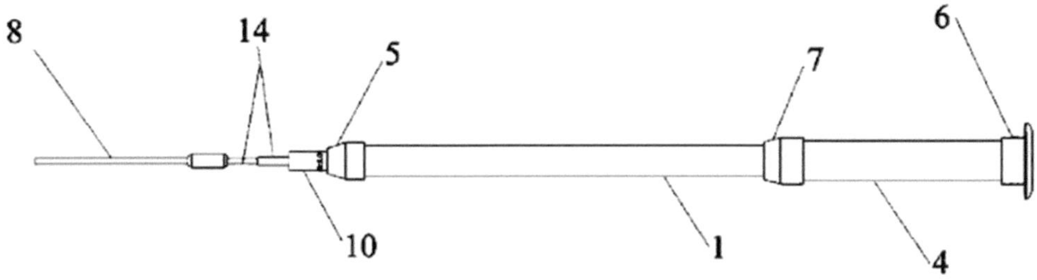

Fig. 1

Fig. 7.6 Figure 1 of DE202020106967U1[13]

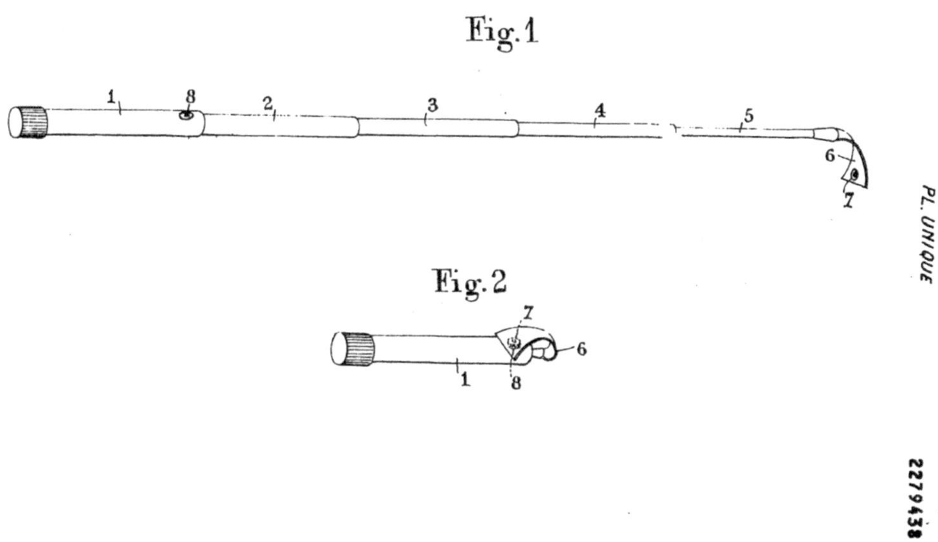

Fig. 7.7 Figures 1 and 2 of FR2279438[15]

[13] DPMA, https://depatisnet.dpma.de/DepatisNet/depatisnet?action=pdf&docid=DE202020106967U1&xxxfull=1, accessed on 29.5.2025.

[15] DPMA, https://depatisnet.dpma.de/DepatisNet/depatisnet?action=pdf&docid=FR000002279438A1&xxxfull=1, accessed on 29.5.2025.

The features of the initially filed main claim of the patent application are therefore fully anticipated by the prior art.

The new claim of the patent application reads as follows:

1. *"Riding crop as equipment for a rider comprising: a handle as a first section (1) and a second section (2), wherein the length of the first section (1) together with that of the second section (2) corresponds to the prescribed length of a riding crop for a first discipline of an equestrian competition,* ***characterized in that****the riding crop comprises a third section (3), wherein the length of the first section (1), together with the second and third sections (2, 3), corresponds to the prescribed length of a riding crop for a second discipline of an equestrian competition, wherein the length of the riding crop can be changed in steps, and wherein an actuation device (4) allows any desired riding crop length to be selected."*[16]

The features following "characterized in that" have been newly added, now including the feature that the respective lengths correspond to the regulations of the respective disciplines. In addition, the new riding crop includes an actuation device 4, with which the desired lengths can be set automatically.

An actuating device 4 and the adjustment of the lengths according to the respective disciplines could not be found in the prior art. The new main claim was therefore novel. Figure 7.8 shows the riding crop according to the invention with the actuating device 4.

The original objective was seen in the fact that the riding crop according to the invention is comfortable to handle. A riding crop that is suitable for different disciplines, in which different lengths of the riding crop are required, can be described as comfortable, since it is convenient to be able to use only one riding crop for different riding disciplines.

The specific function of the riding crop according to the invention is to quickly set the appropriate lengths for the discipline. A general objective is: to provide a riding crop with which different disciplines can be ridden.

7.7 Example 3: Combine Harvester

Patent application DE 10 2020 130 169 A1 describes a combine harvester for harvesting grain. Immediately after harvesting, in a further work step, seeds are sown, which in the prior art usually takes place several weeks later. The objective of the invention is therefore to make the work steps of harvesting and sowing more efficient.

[16] DPMA, https://depatisnet.dpma.de/DepatisNet/depatisnet?action=pdf&docid=DE102022105981A1&xxxfull=1, accessed on 31.12.2024.

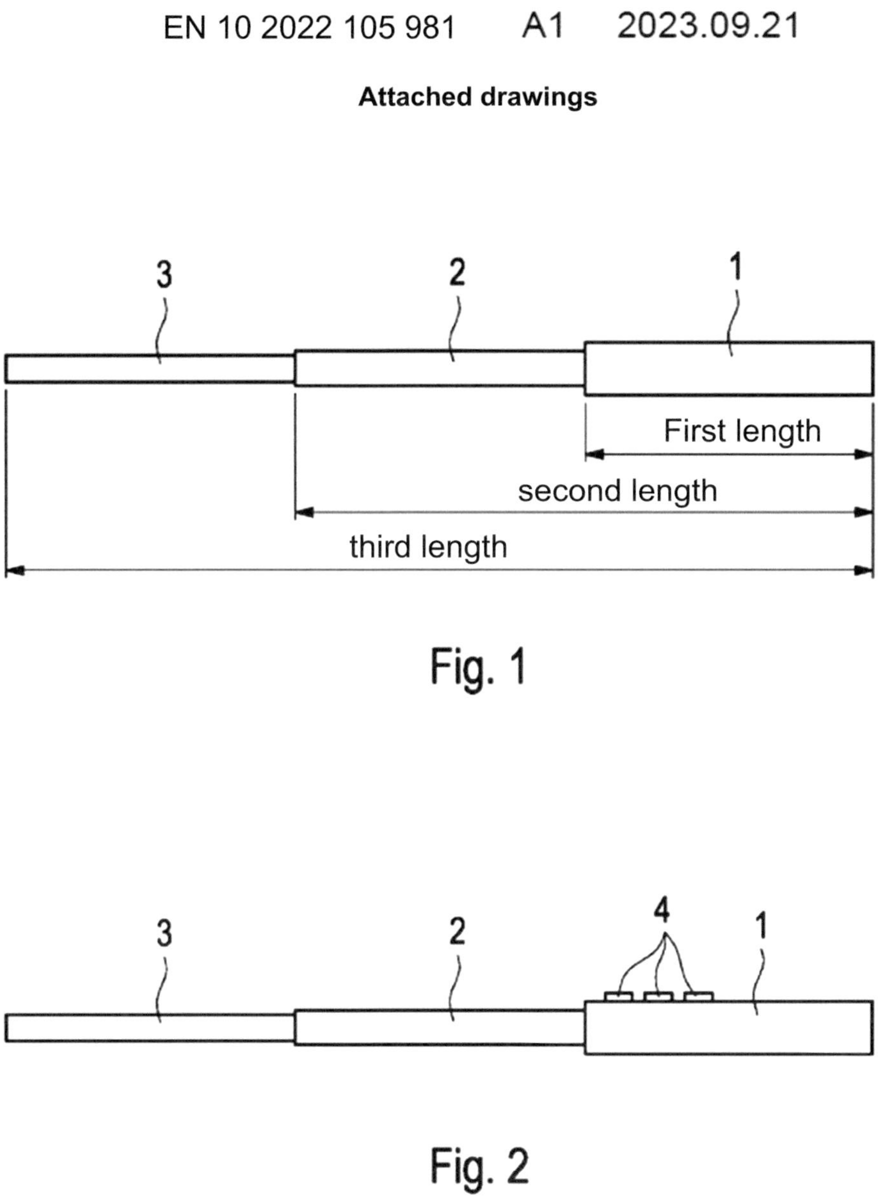

Fig. 7.8 Figures 1 and 2 of DE102022105981A1[17]

[17] DPMA, https://depatisnet.dpma.de/DepatisNet/depatisnet?action=pdf&docid=DE102022105981A1&xxxfull=1, accessed on 29.5.2025.

The objective of the application was initially stated as follows:

"An objective of the invention is therefore to provide a harvesting machine, in particular a combine harvester, which can perform the operations of harvesting and sowing overall more quickly."[18]

The main claim of the application was initially as follows:

1. *"Harvesting machine, in particular combine harvester (1) for cutting crops, comprising:*
   - *a device for cutting (38) crops, in particular cereals such as wheat, rye, oats as well as corn and rapeseed, and*
   - *a seed dispensing device (27) for receiving seed."*[19]

The main claim describes that the harvesting machine comprises a seed dispensing device, so that sowing can take place immediately after harvesting. In this way, the work steps of harvesting and sowing are combined into a single operation.

The patent office identified the prior art documents DE 3644767, US 2016/0212931 A1, and DE 2003879.

In Fig. 7.9, the harvesting machine 1 of the prior art document DE 3644767 is shown, which has a container 5 for receiving seed.[20]

The document US 2016/0212931 also discloses a harvesting machine with a seed dispensing device.

Figure 7.10 shows a harvesting machine 100 from prior art document US 2016/0212931. The harvesting machine 100 of US 2016/0212931 A1 features a cutting unit 12 and a container 16 that holds seed. The seed can be dispensed from nozzle 24 onto the agricultural land after the crops have been cut and harvested.[22]

DE 2003879 also discloses a harvesting machine that features a seed dispensing unit, allowing the same vehicle to be used for sowing the new crop, preferably immediately after harvesting.

Figure 7.11 shows the seed container 10 of DE 2003879, which is arranged on a harvesting machine.[24]

---

[18] DPMA, https://depatisnet.dpma.de/DepatisNet/depatisnet?action=pdf&docid=DE102020130169A1&xxxfull=1, accessed on 4.12.2024.

[19] DPMA, https://depatisnet.dpma.de/DepatisNet/depatisnet?action=pdf&docid=DE102020130169A1&xxxfull=1, accessed on 4.12.2024.

[20] DPMA, https://depatisnet.dpma.de/DepatisNet/depatisnet?action=pdf&docid=DE000003644767A1&xxxfull=1, accessed on 2.1.2025.

[22] DPMA, https://depatisnet.dpma.de/DepatisNet/depatisnet?action=pdf&docid=US020160212931A1&xxxfull=1, accessed on 1/2/2025.

[24] DPMA, https://depatisnet.dpma.de/DepatisNet/depatisnet?action=pdf&docid=DE000002003879A&xxxfull=1, accessed on 1/2/2025.

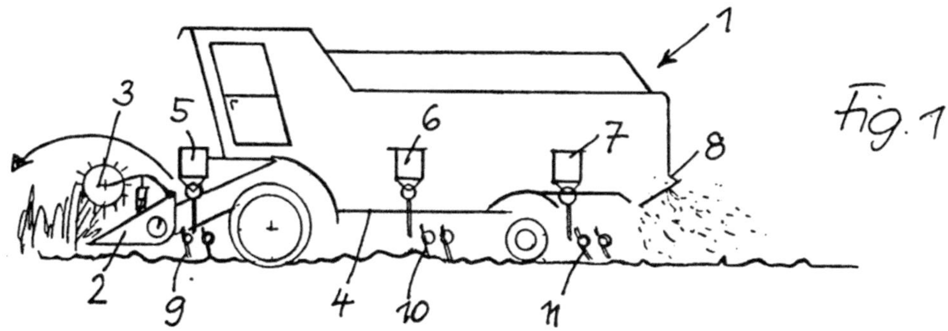

Fig. 7.9 Figure 1 of DE3644767[21]

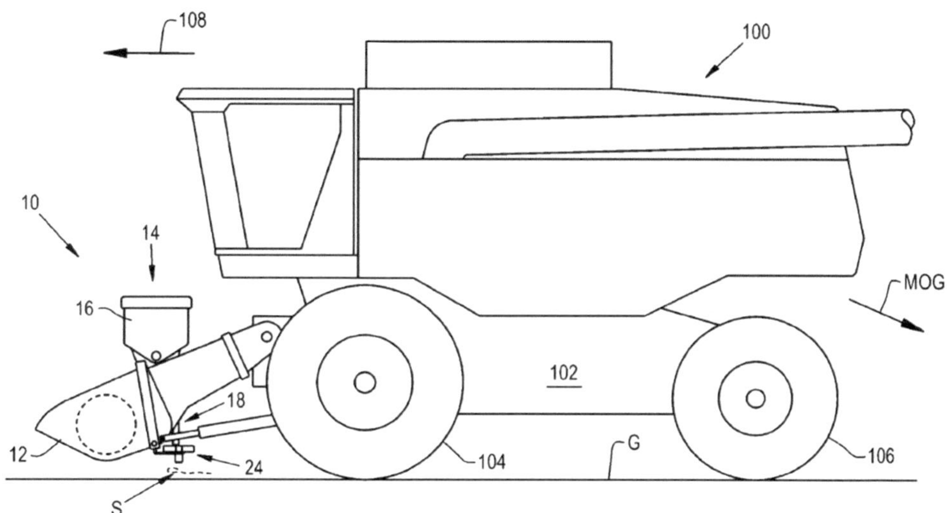

Fig. 7.10 Figure 1 of US20160212931[23]

The main claim of the applicant's own patent application therefore had to be supplemented with additional features to distinguish it from the prior art.

A disadvantage of the seed containers in the prior art documents is the poor visibility for the operator of the harvesting machine, since the containers in the prior art are arranged at the front and top, thereby restricting the operator's view.

---

[21] DPMA, https://depatisnet.dpma.de/DepatisNet/depatisnet?action=pdf&docid=DE000003644767A1&xxxfull=1, accessed on 29.5.2025.

[23] DPMA, https://depatisnet.dpma.de/DepatisNet/depatisnet?action=pdf&docid=US020160212931A1&xxxfull=1, accessed on 5/29/2025.

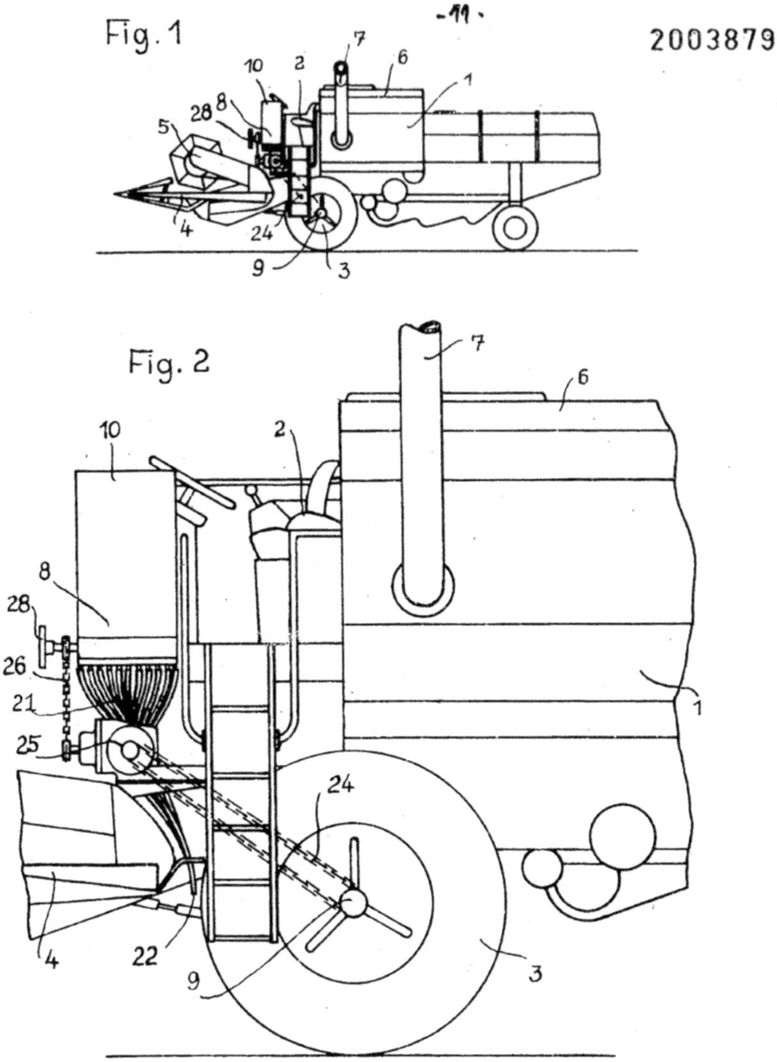

Fig. 7.11 Figures 1 and 2 of DE2003879[25]

A new objective of the application was therefore to arrange the seed dispensing device on the harvesting machine in such a way that the operator's view is not obstructed by the seed dispensing device.

[25] DPMA, https://depatisnet.dpma.de/DepatisNet/depatisnet?action=pdf&docid=DE000002003879A&xxxfull=1, accessed on 5/29/2025.

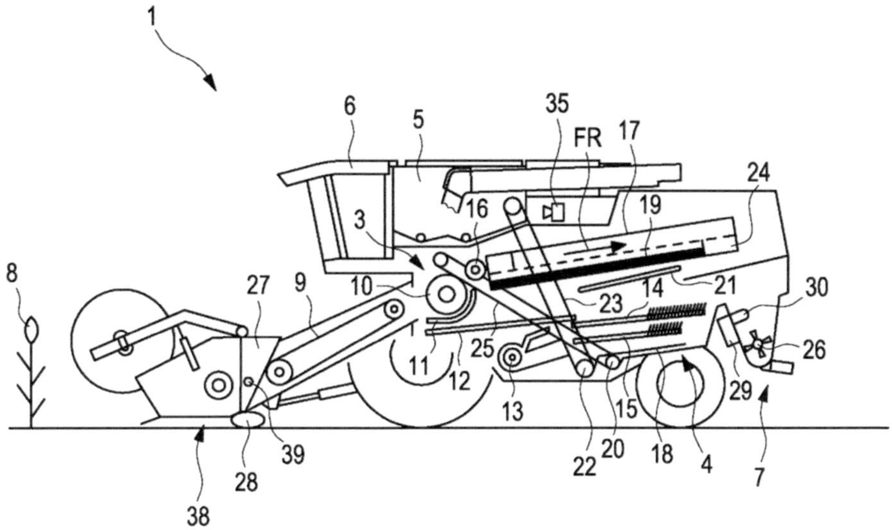

Fig. 1

Fig. 7.12 Figure 1 of DE102020130169B4[27]

The new claim of the applicant's own patent application therefore read as follows:

1. *"Harvesting machine, in particular a combine harvester (1) for cutting crops, comprising:*
   - *a device for cutting (38) crops, in particular cereals such as wheat, rye, oats, as well as corn and rapeseed, and*
   - *a seed dispensing device (27) for receiving seed, wherein the seed dispensing device (27) and the device for cutting (38) share a common wall.*"[26]

The added feature that "the seed dispensing device (27) and the device for cutting (38) share a common wall" can be found in paragraph [0020] of the applicant's own published application.

This feature was used to claim the embodiment shown in Fig. 7.12 of the applicant's own patent application, in which the seed dispensing device 27 is arranged directly behind the cutting unit 38, so that the operator's view from the cab 6 is not obstructed by

[26] DPMA, https://depatisnet.dpma.de/DepatisNet/depatisnet?action=pdf&docid=DE102020130169B4&xxxfull=1, accessed on 12/4/2024.

[27] DPMA, https://depatisnet.dpma.de/DepatisNet/depatisnet?action=pdf&docid=DE102020130169B4&xxxfull=1, accessed on 5/29/2025.

a seed container, as is the case in the prior art documents with container 5 of DE3644767 or container 16 of US20160212931A1.

In Fig. 7.12, the special inventive embodiment is shown, in which the seed dispensing device 27 is arranged directly behind the cutting unit 38 and shares a common wall with it, so that the operator has an unobstructed view to the front.

7.8 Example 4: Fertilizer Device in a Combine Harvester

DE 10 2020 005 450 claims in the patent application a working method for the efficient management of agricultural land.

The original main claim read as follows:

1. *"Working method for a combine harvester (1), comprising the steps:*
   - *Providing a container device (27) for seed (32), a seed dispensing device (28), a container device (29) for*
     liquid fertilizer (34), and a fertilizer dispensing device (30) on the combine harvester (1), furthermore
   - *Carrying out during a single pass the steps:*
     Cutting crops using a cutting unit (2),
     Sowing the seed (32) using the seed dispensing device (28),
     Wetting of chopped material (31) and/or the sown seed (32)
     with liquid fertilizer (34) by means of the fertilizer dispensing device (30), and
     Distributing the chopped material (31) using a distributor (7) over the sown seed (32)."[28]

The original objective of the invention was to prevent the drying out of arable soil. After harvesting the crops, the cut root system of the harvested crops, together with any remaining chopped material, is plowed under, leaving the top layer of arable soil unprotected against drying out and wind erosion.

The patent office was able to identify, in particular, DE 36 44 767 as relevant prior art, which anticipates the working method of the original main claim and destroys its novelty.[29] The subject matter of the main claim was therefore not novel.

However, the set of claims in the patent application included an independent apparatus claim, which read as follows:

[28] DPMA, https://depatisnet.dpma.de/DepatisNet/depatisnet?action=pdf&docid=DE102020005450A1&xxxfull=1, retrieved on 14.10.2024.

[29] DPMA, https://depatisnet.dpma.de/DepatisNet/depatisnet?action=pdf&docid=DE000003644767A1&xxxfull=1, retrieved on 2.1.2025.

7. *"Combine harvester (1), comprising*
   - *a container device (27) for seed (32),*
   - *a seed dispensing device (28) connected to the container device,*
   - *a container device (29) for liquid fertilizer (34), and*
   - *a fertilizer dispensing device (30) connected to the container device (29) for liquid fertilizer, wherein the*
   seed dispensing device is suitable for sowing the seed while the combine harvester (1) is moving and during threshing,
   wherein the fertilizer dispensing device is suitable for wetting the chopped material (31) produced by the combine harvester with liquid fertilizer before distributing it over the sown seed."[30]

In this claim, a fertilizer dispensing device is described that is designed in such a way that the chopped material produced by the combine harvester is mixed with liquid fertilizer in the fertilizer dispensing device and then thrown onto the sown seed. This feature could not be derived from the prior art.

This feature ensures thorough soaking of the chopped material with fertilizer, so that the seed is fertilized very well and evenly, laying the foundation for good seed growth.

A problem statement very close to the invention would be "the optimal fertilization of the seed". A somewhat more remote problem statement is "to promote the growth of the seed", which, of course, can also be achieved by means other than just fertilization.

Accordingly, the objective was amended to:

The objective technical problem is to provide a combine harvester that ensures a high yield of crops.

7.9 Example 5: Shopping Cart with Object Holder

Patent application WO 2021/160772 (EP 21705479) claims an object holder on a shopping cart. The initially filed main claim read as follows:

1. *"Push handle unit (5) of a shopping cart (3) with at least a first and a second end piece (10), wherein at least one end piece (10) is provided with a free-standing, upright gripping unit (11), wherein the gripping unit (11) is designed as an object holder (6) for receiving at least one beverage cup and/or a scanner (1) and/or a magnifying glass for enlarging text on a package and/or a handle cap and/or a smartphone (2)".*[31]

---

[30] DPMA, https://depatisnet.dpma.de/DepatisNet/depatisnet?action=pdf&docid=DE102020005450A1&xxxfull=1, retrieved on 14.10.2024.

[31] EPO, https://register.epo.org/application?documentId=E5VCPJ3P3409DSU&number=EP21705479&lng=de&npl=false, retrieved on 18.10.2024.

The object holder according to the invention on the shopping cart is intended to be suitable for various applications, namely for holding a beverage cup, a scanner, a magnifying glass, and a smartphone. If the receptacle of the object holder is not needed, it can be closed with a cap.

The problem described in the patent application was as follows:

"An object of the invention is therefore to provide a scanner for a shopping cart, with which retrofitting a conventional shopping cart can be carried out quickly and inexpensively."[32]

The initially formulated problem focused on the fact that the object holder accommodates a scanner, with which the customer can independently scan the goods. This can speed up the payment process and the supermarket requires less staff.

Figure 7.13 shows the object holder 6 on a shopping cart 3, into which a scanner or a coffee cup can be inserted.

In this case, unfortunately, the applicant herself had previously filed a similar patent application in which Fig. 3 of Fig. 7.13 was depicted.

It was therefore necessary to formulate a new main claim:

1. *"Push handle unit (5) of a shopping cart (3) with at least a first and a second end piece (10), wherein at least one end piece (10) is provided with a free-standing, upright gripping unit (11), wherein the gripping unit (11) is designed as an object holder (6) for receiving a magnifying glass for enlarging text on a package and a smartphone (2), characterized in that the smartphone (2) can be connected to the object holder (6) and can thus be used as a mobile scanner."*[34]

Through the inventive features of the possibility to connect a smartphone to the object holder, the customer's smartphone can be used as a scanner. The technical problem was therefore to enable the use of a smartphone during a shopping process, with the object holder providing the appropriate connection options.

7.10 Example 6: Push Handle of a Shopping Cart

WO 2022/089792 A1 describes a push handle of a shopping cart, which is composed of a handle bar and two handle caps, with a detachable connection between the handle caps and the handle bar enabled by clipping them in. The advantage of this design is the rapid assembly of the push handle onto a shopping cart, allowing for cost-effective manufacturing.

[32] EPO, https://register.epo.org/application?documentId=E5VCPJLR6714DSU&number=EP21705479&lng=de&npl=false, retrieved on 19.10.2024.

[34] DPMA, https://depatisnet.dpma.de/DepatisNet/depatisnet?action=pdf&docid=EP000004041614B1&xxxfull=1, retrieved on 31.5.2025.

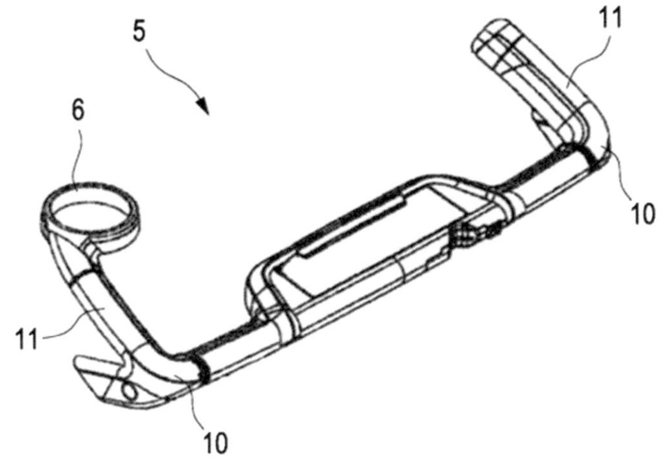

Fig. 3

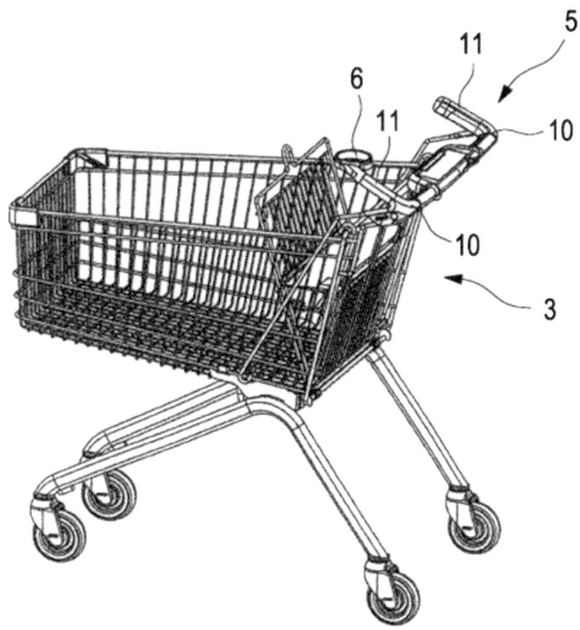

Fig. 4

Fig. 7.13 Figures 3 and 4 of EP21705479[33]

---

[33] EPO, https://register.epo.org/application?number=EP21705479, retrieved on 29.5.2025.

Fig. 8

Fig. 7.14 Figure 8 of WO2022089792A1[36]

The application states the following as the technical problem:

"*Typically, the handle caps are screwed onto the handle bar. Assembly of a push handle unit is therefore time-consuming.*

An object of the invention is therefore to provide a handle cap, a push handle unit, and a shopping cart, so that the assembly of a handle cap onto a handle bar can be carried out quickly in terms of time."[35]

Figure 7.14 shows a shopping cart with an inventive handle bar 6 and handle caps 2.

In Fig. 7.15, the handle cap 2 can be seen, which engages with the handle bracket 6 via spring elements 1 into the wire loop 5 of the shopping cart, thereby creating, according to the invention, a detachable connection.

[35] DPMA, https://depatisnet.dpma.de/DepatisNet/depatisnet?action=pdf&docid=WO002022089792A1&xxxfull=1, retrieved on 2.12.2024.

[36] DPMA, https://depatisnet.dpma.de/DepatisNet/depatisnet?action=pdf&docid=WO002022089792A1&xxxfull=1, retrieved on 29.5.2025.

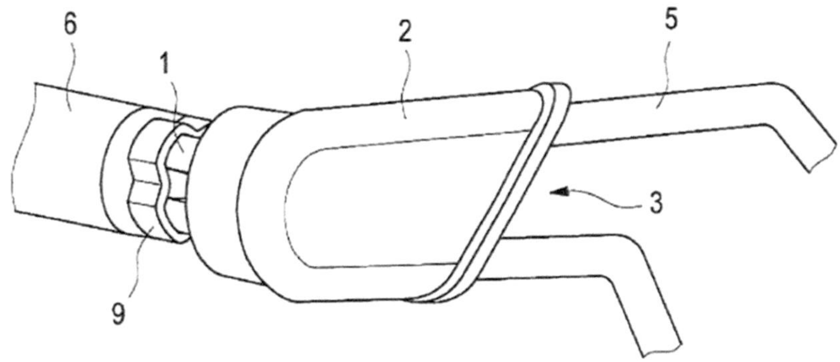

Fig. 7.15 Figure 10 of WO2022089792A1[37]

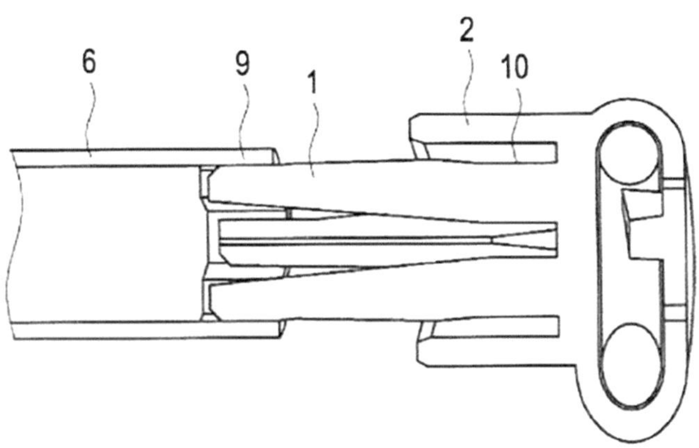

Fig. 7.16 Figure 13 of WO2022089792A1[38]

Figure 7.16 shows the spring elements 1 of the handle cap 2, which establish the connection with the handle bracket 6.

The European Patent Office identified GB 2014527 A, DE 3044581 A1, and DE 19830297 A1 as relevant prior art.

[37] DPMA, https://depatisnet.dpma.de/DepatisNet/depatisnet?action=pdf&docid=WO002022089792A1&xxxfull=1, accessed on 29.5.2025.

[38] DPMA, https://depatisnet.dpma.de/DepatisNet/depatisnet?action=pdf&docid=WO002022089792A1&xxxfull=1, accessed on 29.5.2025.

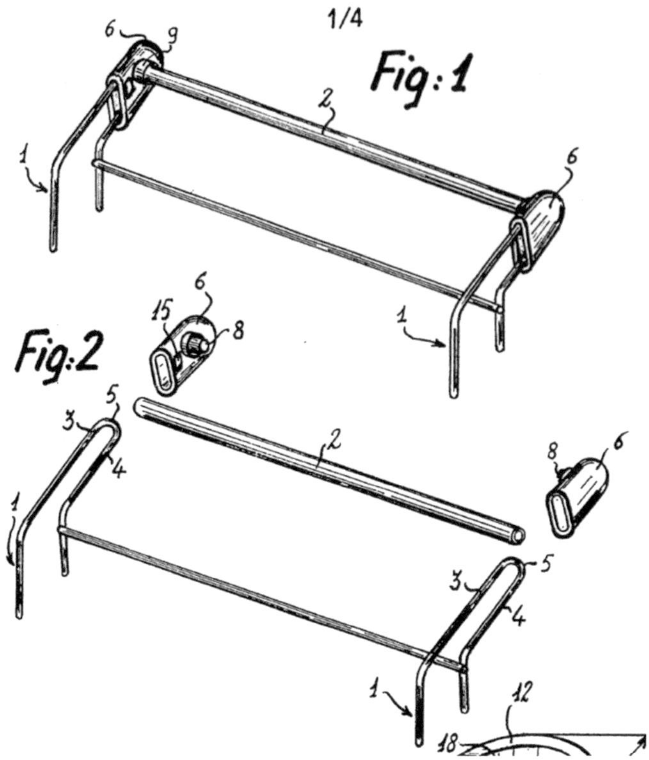

Fig. 7.17 Figs. 1 and 2 of GB2014527A[40]

In Fig. 7.17, the handle caps 6 and the handle bracket 2 of the prior art GB 2014527 are shown, which are also designed to be assembled together.[39]

Figure 7.18 shows that a bolt 8 engages in the handle cap 6 and thus, via the wire loop, creates a detachable connection.

The second prior art reference, DE 3044581 A1, shows a shopping cart with a push handle 7, whereby the technical teaching of the document describes the connection of the complete push handle unit 7 with the shopping cart.[42]

---

[39] DPMA, https://depatisnet.dpma.de/DepatisNet/depatisnet?action=pdf&docid=GB000002014527A&xxxfull=1, accessed on 2.1.2025.

[40] DPMA, https://depatisnet.dpma.de/DepatisNet/depatisnet?action=pdf&docid=GB000002014527A&xxxfull=1, accessed on 29.5.2025.

[42] DPMA, https://depatisnet.dpma.de/DepatisNet/depatisnet?action=pdf&docid=DE000003044581A1&xxxfull=1, accessed on 2.1.2025.

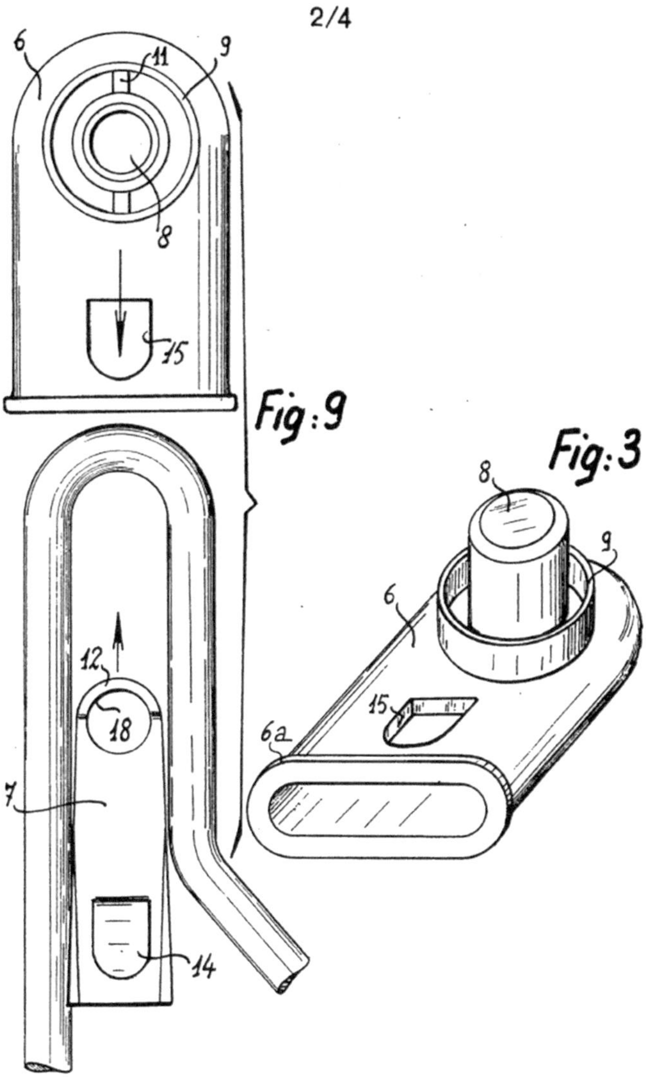

Fig. 7.18 Figures 3 and 9 of GB2014527A[41]

Figure 7.19 shows the technical teaching of the prior art DE 3044581 A1, which deals with the connection of the push handle unit 7 with the shopping cart and not with the connection of handle cap 6 and handle bracket 7. This document is therefore irrelevant.

[41] DPMA, https://depatisnet.dpma.de/DepatisNet/depatisnet?action=pdf&docid=GB00000201452 7A&xxxfull=1, accessed on 29.5.2025.

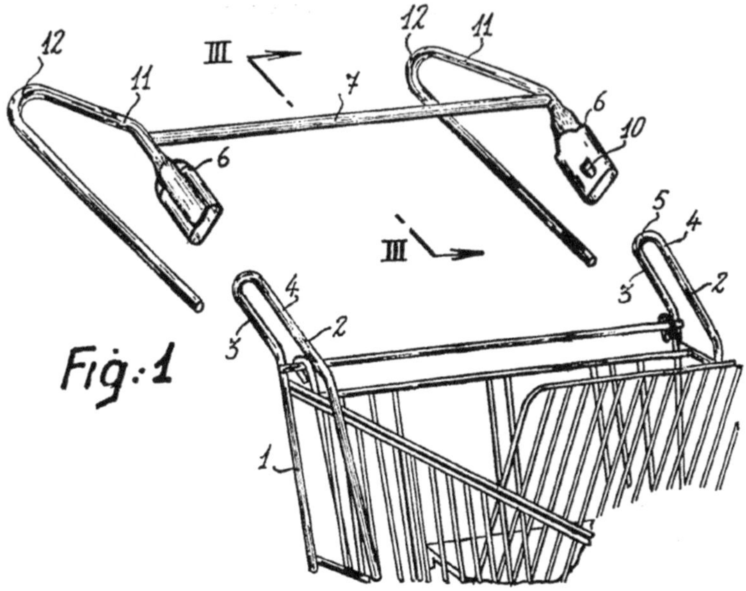

Fig. 7.19 Figure 1 of DE3044581A1[43]

The third prior art document DE 198 30 297 A1 describes the connection of a handle cap with a handle bracket, whereby a projection of the handle cap can be inserted into a complementary recess of the handle bracket.[44]

Figure 7.20 shows a handle cap 11 from prior art DE 198 30 297 A1, which enables a detachable connection with a handle bracket 2.

The problem addressed by the invention had therefore already been solved in the prior art in the manner described by the applicant's own technical teaching. It was therefore necessary to formulate a new problem in order to arrive at a technical teaching that is not contained in the prior art and is not suggested by it.

To distinguish from the prior art, the feature could be used that several spring elements engage in a wire loop, thereby creating the detachable connection.

Figure 7.21 shows several spring elements 1 of the handle cap 2 of the applicant's own invention, which are clipped into a wire loop of the handle bracket to create a detachable connection between the handle bracket and the handle cap 2. This results in a detachable connection in which only one element, namely the wire loop, is present on one side of the connection, and several elements, namely the spring elements 1, are present on the other side. This particular embodiment of the detachable connection allows

---

[43] DPMA, https://depatisnet.dpma.de/DepatisNet/depatisnet?action=pdf&docid=DE000003044581A1&xxxfull=1, accessed on 29.5.2025.

[44] DPMA, https://depatisnet.dpma.de/DepatisNet/depatisnet?action=pdf&docid=DE000019830297A1&xxxfull=1, accessed on 2.1.2025.

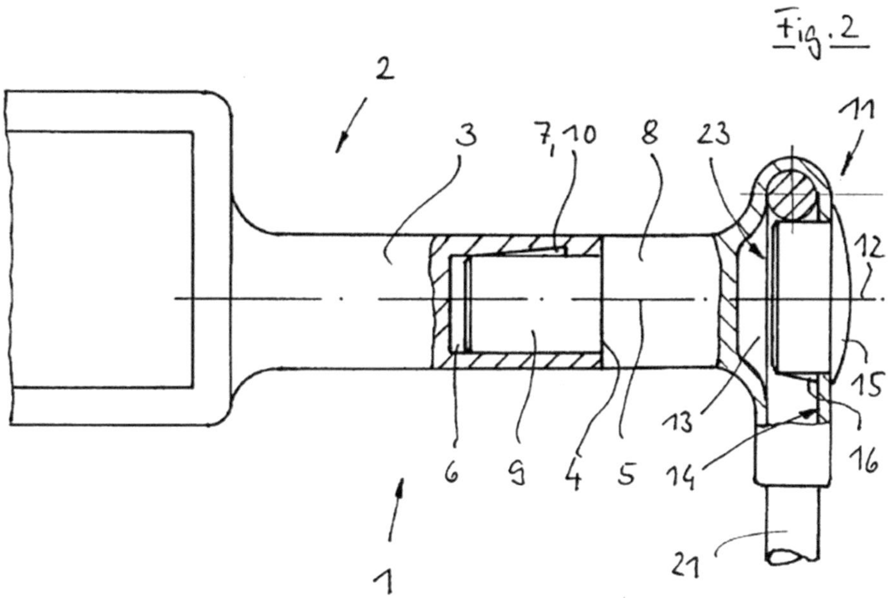

Fig. 7.20 Figure 2 of DE19830297A1[45]

for a very compact connection, with several elements involved on one side of the connection, thus enabling a very stable connection at least on that side.

The new objective technical problem was therefore to provide a detachable connection of a handle cap with a handle bracket of a shopping cart, which is compact and at least on one side of the connection is mechanically very stable.

The new main claim of the applicant's own patent application was as follows:

1. *"Handle unit (2, 6) for arrangement on a shopping cart, comprising a handle bracket (6) and a handle cap (2) as a lateral end of the handle unit of the shopping cart, which establishes the connection of the handle bracket (6) with the shopping cart, wherein the handle cap (2) has a first element that enables a detachable or non-detachable connection with a second element of the handle bracket (6), characterized in that the first element comprises 2, 3, 4, 5 or any number of spring elements (1) and the second element is a wire loop, or wherein the first element is a wire loop and the second element comprises 2, 3, 4, 5 or any number of spring elements (1)."*[47]

[45] DPMA, https://depatisnet.dpma.de/DepatisNet/depatisnet?action=pdf&docid=DE000019830297A1&xxxfull=1, accessed on 29.5.2025.

[47] EPA, https://register.epo.org/application?number=EP21740503&lng=en&tab=doclist, accessed on 3.12.2024.

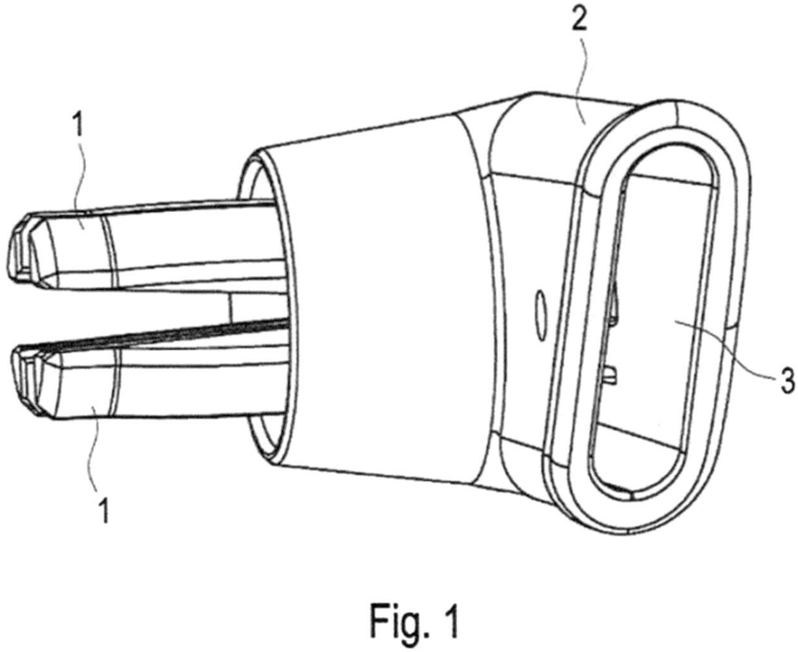

Fig. 7.21 Figure 1 of WO2022089792A1[46]

7.11 Example 7: Locking device for a shopping cart

Coin deposit systems, in which, for example, a 1-euro coin is inserted to release a shopping cart from a chain, have become widely established. Document WO 2023/006962 (EP22760679, EP4377863) describes a coin deposit system for a shopping cart, in which the customer can release a shopping cart not only with a coin but alternatively with a signal from a smartphone. Thus, a hybrid locking device for a shopping cart is provided.

The initial objective of the invention was as follows:

"An object of the invention is therefore to provide a device and a method that offers an additional option for releasing a shopping cart from a chain of shopping carts, or provides an alternative way to release a shopping cart from a group of shopping carts without providing a deposit."[48]

[46] DPMA, https://depatisnet.dpma.de/DepatisNet/depatisnet?action=pdf&docid=WO002022089792A1&xxxfull=1, accessed on 29.5.2025.

[48] EPA, https://register.epo.org/application?number=EP22760679&lng=de&tab=doclist, accessed on 4.12.2024.

The problem was solved with the following main claim:

1. *"Hybrid locking device (11) of a transport cart (22), in particular a shopping cart, for detachably connecting with a key (3) of another transport cart (22), wherein the locking device (11) is suitable for at least partial reception of the key (3),*
   characterized in that
   the locking device (11) releases a received key (3) by inserting a coin (9) and/or by an electrical or electromagnetic signal."[49]

Figure 7.22 shows the inventive locking device with an inserted coin 9, which is gripped by gripping arms when the key or pin 3 is removed from the locking device.[50]

The patent office researched the prior art documents DE 10 2017 001 920 A1, AU 2010 100 636 A4, and CN 108 824 997 A.

Figure 7.23 shows a locking system from the prior art DE 10 2017 001 920 A1 with a smartphone 3, which is used to unlock a shopping cart.[52]

Figure 7.24 shows a conventional coin deposit system according to the prior art AU 2010100636 A4, in which the insertion of a coin 2 into the locking device releases the key or pin 3.[54]

The prior art CN 108 824 997 A discloses a hybrid device in which the key 120 can be unlocked either with a coin or with an electromagnetic signal originating from a smartphone.

Figure 7.25 shows a locking device 110 with a slot for a coin 112 from the prior art CN 108 824 997 A, for releasing the key 120 from a chain of shopping carts.[56]

Figure 7.26 shows the internal structure of the locking device of prior art CN 108 824 997 A with a hybrid device that releases the key either by inserting a coin or by a signal from a smartphone.

Prior art CN 108 824 997 A thus discloses a subject matter corresponding to the main claim of the applicant's own patent application. However, CN 108824997 A exhibits a significant distinguishing feature compared to the applicant's invention, namely that in CN 108824997 A a receptacle is laterally extended, into which a coin can be inserted and

[49] EPA, https://register.epo.org/application?number=EP22760679&lng=de&tab=doclist, accessed on 4.12.2024.

[50] EPA, https://register.epo.org/application?documentId=L66E9935VEBEQOJ&number=EP22760679&lng=de&npl=false, accessed on 5.12.2024.

[52] DPMA, https://depatisnet.dpma.de/DepatisNet/depatisnet?action=pdf&docid=DE102017001920A1&xxxfull=1, accessed on 2.1.2025.

[54] DPMA, https://depatisnet.dpma.de/DepatisNet/depatisnet?action=pdf&docid=AU002010100636A4&xxxfull=1, accessed on 2.1.2025.

[56] DPMA, https://depatisnet.dpma.de/DepatisNet/depatisnet?action=pdf&docid=CN000108824997A&xxxfull=1, accessed on 2.1.2025.

Fig. 7.22 Fig. 6 of WO2023/006962[51]

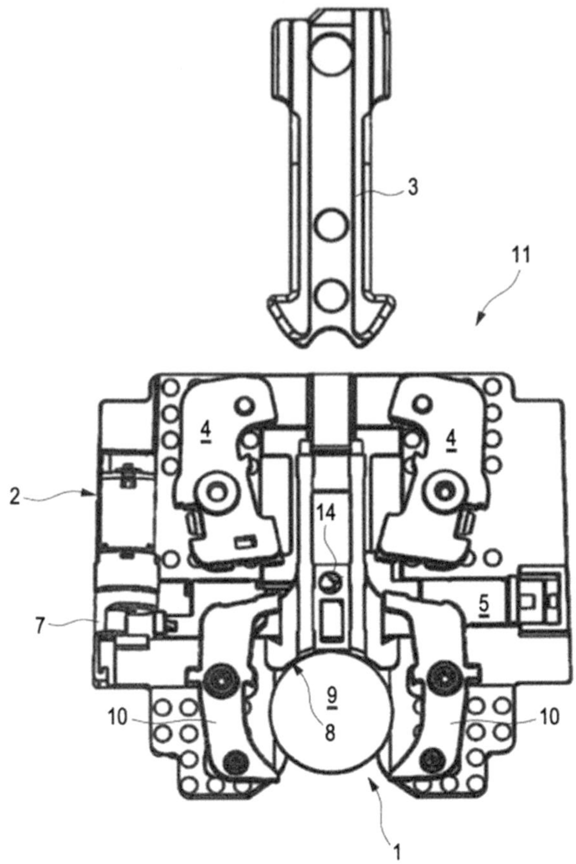

Fig. 6

which must then be pushed back into the device. As a result, this receptacle protrudes unprotected when not in use, which poses a risk of damage to the locking device.

The locking device according to prior art AU 2010100636 A4 exhibits significantly more structural similarities to the present invention, as it features gripping arms into which a coin is inserted to release the key. AU 2010100636 A4 can therefore be made the

[51] DPMA, https://depatisnet.dpma.de/DepatisNet/depatisnet?action=pdf&docid=WO002023006962A1&xxxfull=1, accessed on 29.5.2025.

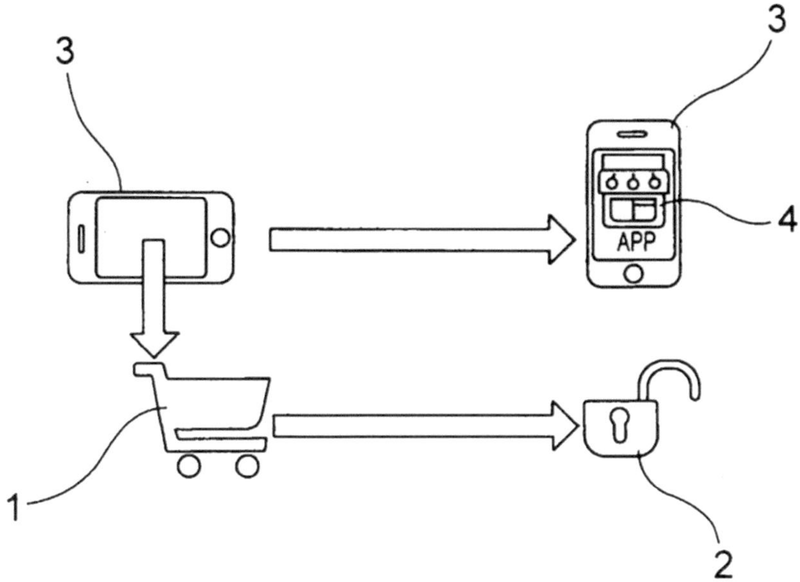

Fig. 7.23 Figure from DE102017001920A1[53]

closest prior art by incorporating the gripping arms into the main claim of the applicant's own patent application. Furthermore, the particular advantage of compactness and the reduction of the risk of damage from mechanical influences should be described in order to distinguish from the present prior art.

The new objective is therefore:

"*The objective technical problem is therefore to provide a locking device for a shopping cart that can be actuated mechanically by coin insertion and by another variant, in particular a smartphone*."[59]

The new main claim incorporates the gripping arms of the device as features. In addition, the hybrid character is emphasized by further features of actuation via smartphone:

1. "*Hybrid locking device (11) of a transport cart (22), in particular a shopping cart, for releasably connecting with a key (3) of another transport cart (22), wherein the locking device (11) is suitable for at least partial reception of the key (3), wherein*

[53] DPMA, https://depatisnet.dpma.de/DepatisNet/depatisnet?action=pdf&docid=DE102017001920A1&xxxfull=1, accessed on 29.5.2025.

[59] EPA, https://register.epo.org/application?number=EP22760679&lng=de&tab=doclist, accessed on 5.12.2024.

Fig. 7.24 Figure 1 of AU2010100636A4[55]

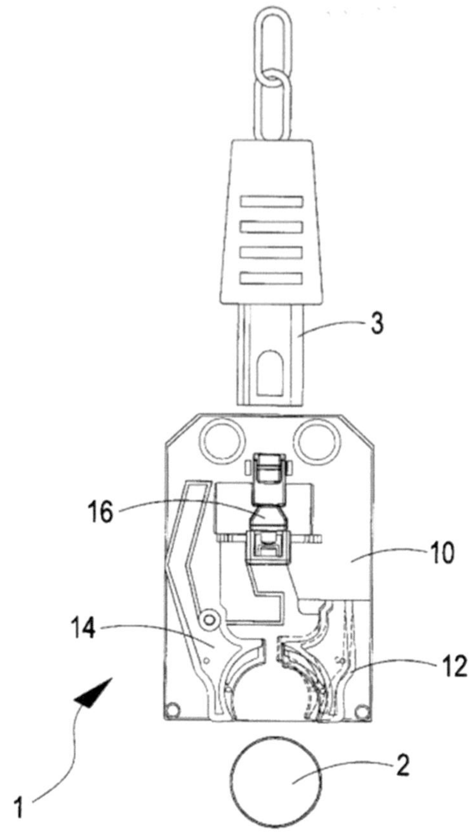

FIG.1

the locking device (11) releases a received key (3) by the insertion of a coin (9) or by an electrical or electromagnetic signal, characterized in that the locking device comprises:
two first gripping arms (4) for locking the key (3),
two second gripping arms (10) for locking the coin (9), wherein the insertion of the coin (9) into the locking device (11) results in a release movement of a slider (5),

[55] DPMA, https://depatisnet.dpma.de/DepatisNet/depatisnet?action=pdf&docid=AU002010100636A4&xxxfull=1, accessed on 29.5.2025.

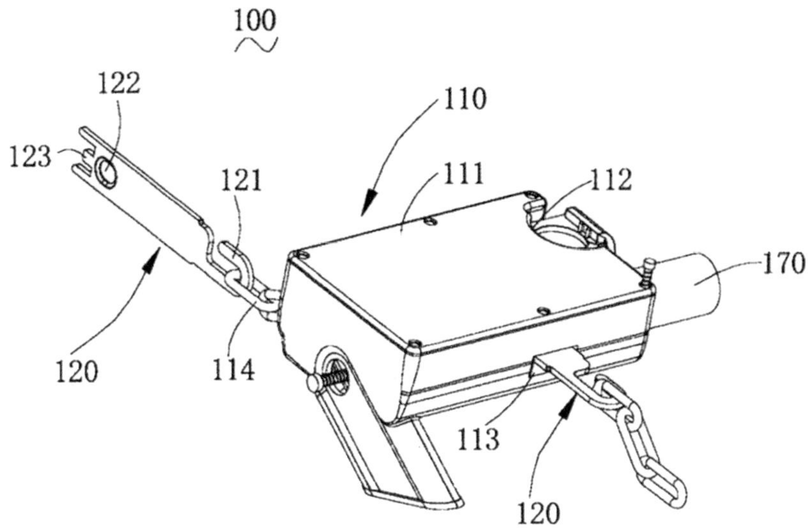

Fig. 7.25 Figure 1 of CN108824997A[57]

> *an electric motor (2) and/or an actuator for driving an eccentric disk (7) to initiate a release movement of the slider (5) and a receiving unit for communication with a transmitting unit, in particular a smartphone (18), for example via Bluetooth, Near Field Communication, or WLAN, wherein the receiving unit controls the motor (2) via the electrical or electromagnetic signal, wherein*

the release movement of the slider (5) opens the gripping arms (4) and thus releases the key (3), wherein the two first and the two second gripping arms (4, 10) perform rotational movements."[60]

[57] DPMA, https://depatisnet.dpma.de/DepatisNet/depatisnet?action=pdf&docid=CN000108824997A&xxxfull=1, accessed on 29.5.2025.

[60] EPA, https://register.epo.org/application?documentId=M3O7RI32IAVKKJ5&number=EP22760679&lng=de&npl=false, accessed on 5.12.2024.

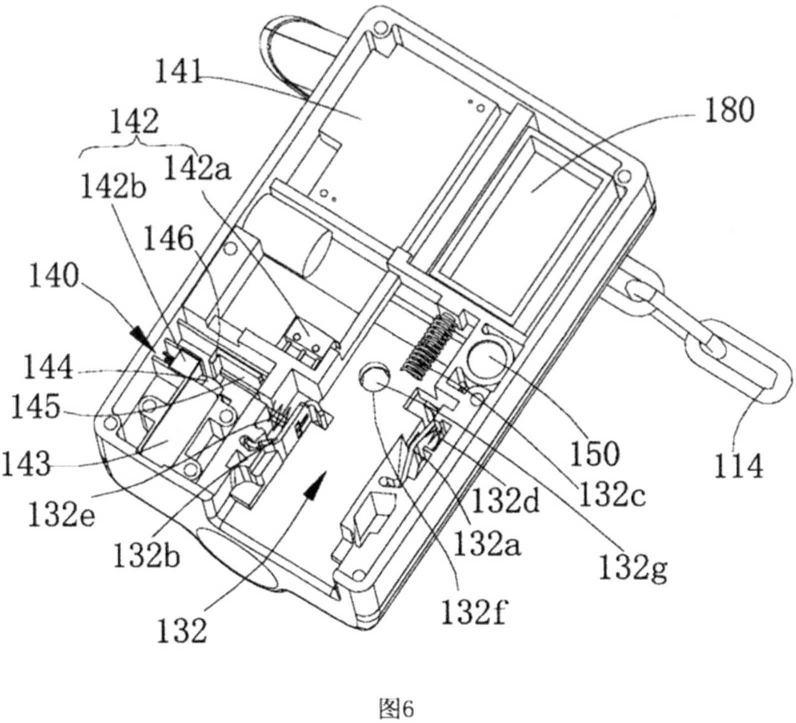

Fig. 7.26 Figure 6 of CN108824997A[58]

7.12 Example 8: Plastic Emblems

A patent application was filed at the patent office in the technical field of fastening emblems to bags, furniture, and vehicle seats. The relevant prior art was DE 10 2012 109 955 A1, which describes an emblem fastening in which the emblem is attached to the carrier element, for example a bag or a piece of furniture, by means of bendable pins.[61]

Figure 7.27 shows an emblem from prior art DE 10 2012 109 955 A1 with pins 17 that are pierced through the carrier element and received in grooves by complementary elements 14, so that the emblem is attached to the carrier element, the bag, the piece of furniture, or the vehicle seat.[62]

[58] DPMA, https://depatisnet.dpma.de/DepatisNet/depatisnet?action=pdf&docid=CN000108824997A&xxxfull=1, accessed on 29.5.2025.

[61] DPMA, https://depatisnet.dpma.de/DepatisNet/depatisnet?action=pdf&docid=DE102012109955A1&xxxfull=1, accessed on 28.12.2024.

[62] DPMA, https://depatisnet.dpma.de/DepatisNet/depatisnet?action=pdf&docid=DE102012109955A1&xxxfull=1, accessed on 28.12.2024.

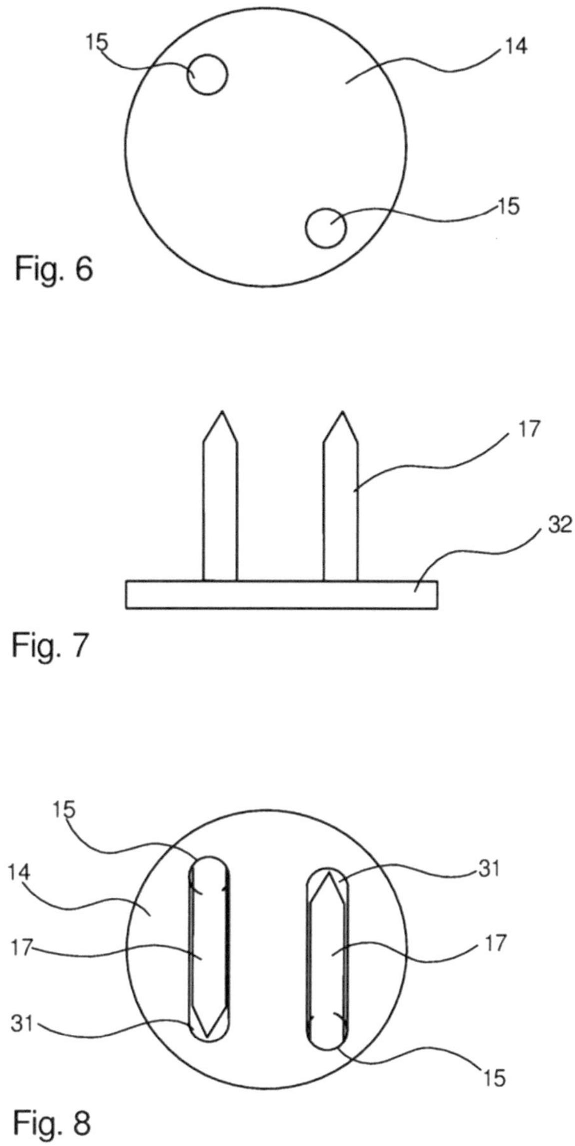

Fig. 7.27 Figs. 6, 7, and 8 of DE102012109955A1[63]

[63] DPMA, https://depatisnet.dpma.de/DepatisNet/depatisnet?action=pdf&docid=DE102012109955A1&xxxfull=1, accessed on 29.5.2025.

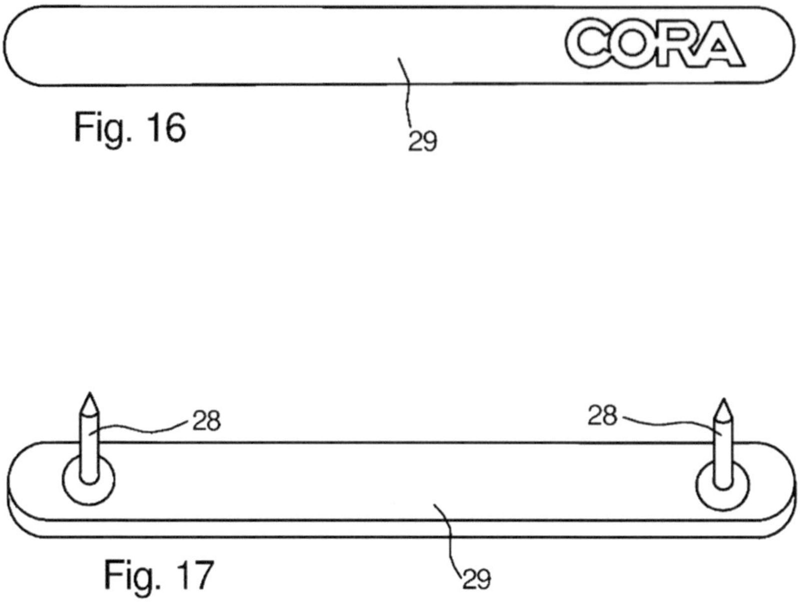

Fig. 7.28 Figures 16 and 17 of DE102012109955A1[64]

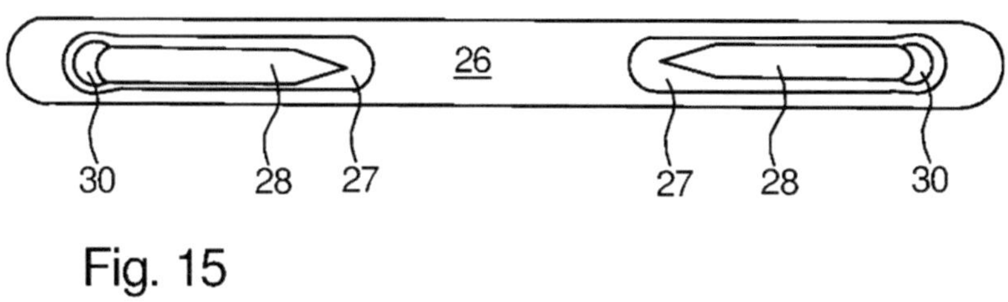

Fig. 7.29 Figure 15 of DE102012109955A1[65]

Figures 7.27, 7.28, and 7.29 show emblems 14 and 29 with pins 28 from the prior art, where the pins 28 are pierced through the carrier element, the bag, the piece of furniture, or the vehicle seat and then bent in such a way that they are received in the grooves 27 of a counterplate 26.

There was a need to prefabricate emblems as much as possible so that the placement of the special mark could be carried out quickly and flexibly. With such inventive

[64] DPMA, https://depatisnet.dpma.de/DepatisNet/depatisnet?action=pdf&docid=DE102012109955A1&xxxfull=1, accessed on 29.5.2025.

[65] DPMA, https://depatisnet.dpma.de/DepatisNet/depatisnet?action=pdf&docid=DE102012109955A1&xxxfull=1, accessed on 29.5.2025.

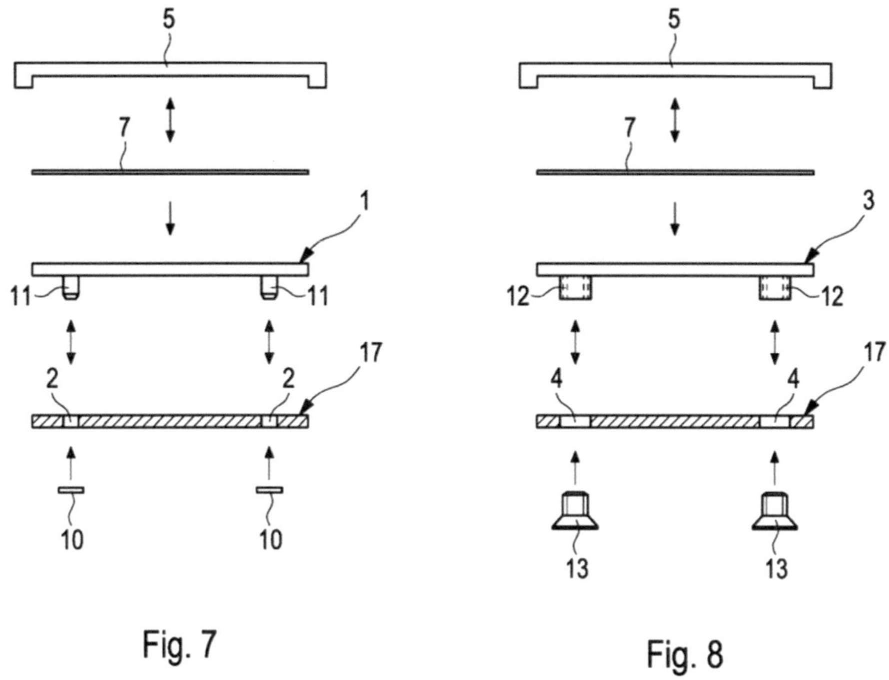

Fig. 7.30 Figures 7 and 8 of DE102019104972B3[67]

devices, the relevant bag, piece of furniture, or car seat can be largely completed, and the attachment of the specific emblem with the desired brand designation can be carried out at short notice.

The objective technical problem was therefore to provide a carrier system for an emblem that is largely pre-assembled, so that final assembly can be carried out flexibly and quickly.

The developed invention was described in the applicant's own patent DE 10 2019 104 972 B3, in which a carrier system is first arranged on the bag in order to attach a customer-specific emblem shortly before delivery.

Figure 7.30 shows the inventive emblem holder, in which a carrier system with a front plate 1 and a rear plate 17 with elements 11 and 10 is attached to a bag, a piece of furniture, or a car seat, and at a later time the emblem 7 is attached to the front plate by means of a cover 5.[66]

---

[66] DPMA, https://depatisnet.dpma.de/DepatisNet/depatisnet?window=1&space=main&content=captcha&action=pdf&docid=DE102019104972B3&xxxfull=1, accessed on 28.12.2024.

[67] DPMA, https://depatisnet.dpma.de/DepatisNet/depatisnet?action=pdf&docid=DE102019104972B3&xxxfull=1, accessed on 29.5.2025.

7.13 Example 9: Physiotherapeutic Device for Rehabilitation

The applicant's own patent EP 18778844.3 (DE 10 2017 122 295 A1) describes a physiotherapeutic device that enables people to relearn movement sequences after brain injury.

The initially submitted main claim read as follows:

1. *"Physiotherapeutic device for treating a patient through physical movement, comprising:*
   an arm trainer (2) for rotational movement of the patient's arms with a first axis, and a leg trainer (1) for rotational movement of the patient's legs with a second axis, wherein the first and second axes are height-adjustable,
   characterized in that
   the distance between the first axis and the second axis is constant."[68]

The description explains:

"Physiotherapeutic devices are used for the rehabilitation of people with illnesses. Such devices are advantageously characterized by the fact that different parts of the body can be trained with a single device. In addition, the device should be easy to operate so that the patient can operate or adjust it themselves, or so that a trained professional, such as a nurse or caregiver, can quickly reconfigure the device or apparatus."[69]

From this, the technical problem was derived:

"An object, therefore, is to provide a device that is characterized by high variability and ease of operation."[70]

The German Patent Office identified the prior art documents US 2016/0206915 A1 and US 5785631, both of which disclose an arm and leg trainer.

Figure 7.31 shows the arm and leg trainer of prior art document US 2016/0206915 A1.[71]

The second prior art document is US 5785631, which shows that converting the device from a sitting to a lying position requires a complex setup process on the treatment device.

[68] DPMA, https://depatisnet.dpma.de/DepatisNet/depatisnet?action=pdf&docid=DE102017122295A1&xxxfull=1, accessed on 19.10.2024.

[69] DPMA, https://depatisnet.dpma.de/DepatisNet/depatisnet?action=pdf&docid=DE102017122295A1&xxxfull=1, accessed on 19.10.2024.

[70] DPMA, https://depatisnet.dpma.de/DepatisNet/depatisnet?action=pdf&docid=DE102017122295A1&xxxfull=1, accessed on 19.10.2024.

[71] DPMA, https://depatisnet.dpma.de/DepatisNet/depatisnet?action=pdf&docid=US020160206915A1&xxxfull=1, accessed on 2.1.2025.

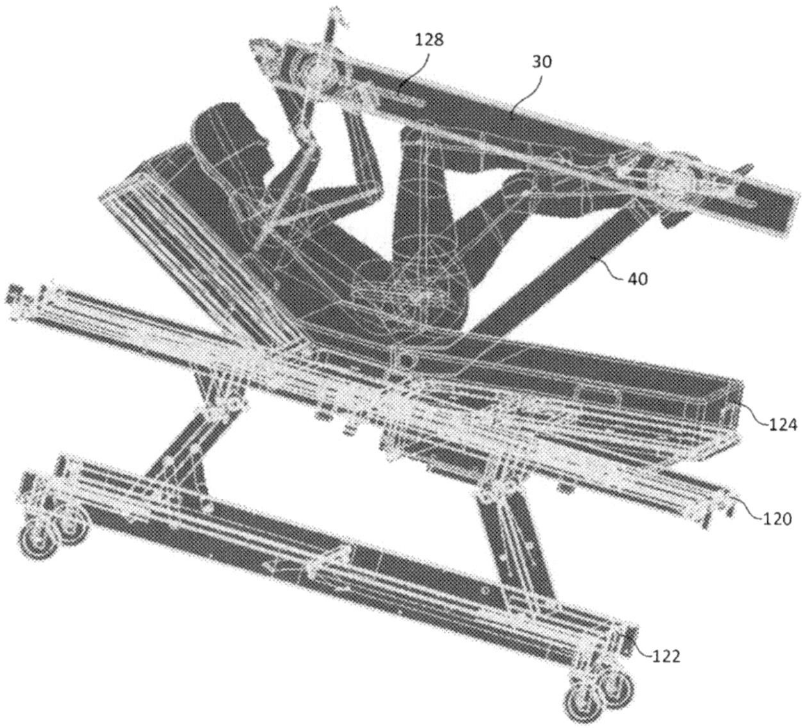

Fig. 7.31 Figure 10 of US20160206915A1[72]

Figures 7.32 and 7.33 show the different training positions of the training device disclosed in US 5785631.[73]

The main claim was therefore amended to read:

1. *"Physiotherapeutic device for treating a patient through physical movement, comprising:*
   an arm trainer (2) for rotational movement of the patient's arms with a first axis, and a leg trainer (1) for rotational movement of the patient's legs with a second axis, wherein the first and second axes are height-adjustable,
   wherein the distance between the first axis and the second axis is constant, wherein the height adjustability of the arm and leg trainers (1, 2) is provided such that, when the height is adjusted, the arm trainer (2) and the leg trainer (1) move uniformly

---

[72] DPMA, https://depatisnet.dpma.de/DepatisNet/depatisnet?action=pdf&docid=US020160206915A1&xxxfull=1, accessed on 29.5.2025.

[73] DPMA, https://depatisnet.dpma.de/DepatisNet/depatisnet?action=pdf&docid=US000005785631A&xxxfull=1, accessed on 2.1.2025.

FIG. 1

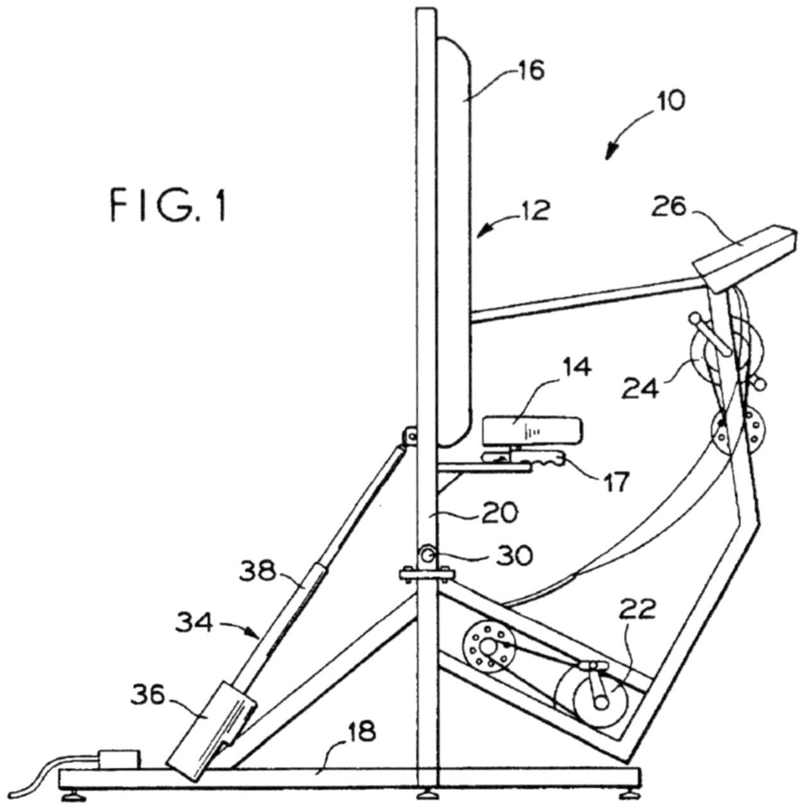

Fig. 7.32 Figure 1 of US5785631[74]

together, so that a different training position of the patient can be achieved, wherein the training position changes from a sitting to a lying position, wherein the device comprises a foot element (10) and a body element (11), with the arm trainer (2) and the leg trainer (1) arranged on the body element (11), ***characterized in that*** *the height adjustability is achieved by shifting the foot element (10) relative to the body element (11).*"[76]

The advantage of the device according to the invention is that the adjustment can be carried out very easily, which enables the patient to convert the device independently or

[74] DPMA, https://depatisnet.dpma.de/DepatisNet/depatisnet?action=pdf&docid=US000005785631A&xxxfull=1, accessed on 29.5.2025.

[76] EPO, https://register.epo.org/application?documentId=LR94J1HMBEML2Y2&number=EP18778844&lng=de&npl=false, accessed on 22.10.2024.

FIG.3

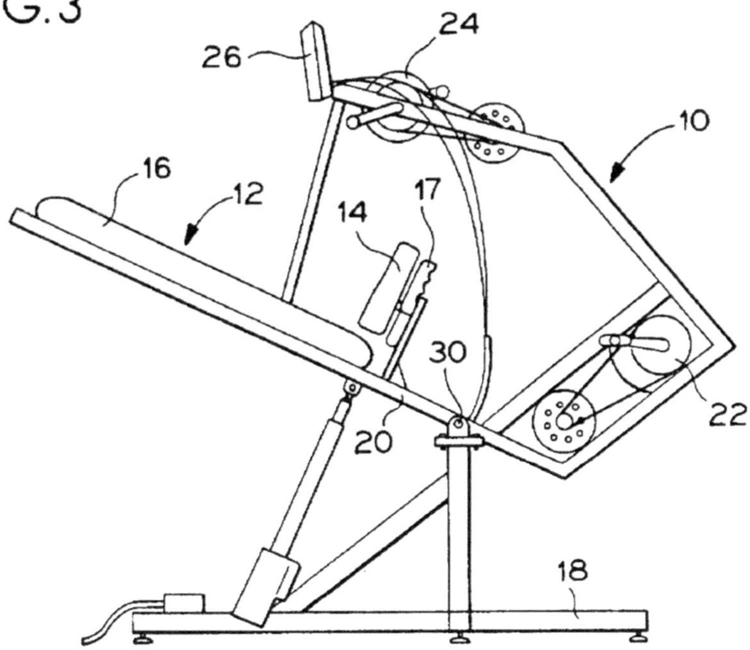

Fig. 7.33 Figure 3 of US5785631[75]

makes it easier for the nursing staff to set up the treatment devices. Another advantage is that the treatment device has a defined space requirement and does not require additional space when being converted.

The new objective was formulated as providing a device that efficiently utilizes a predetermined spatial volume to enable both a sitting and a lying training position.[77]

Figure 7.34 shows the device according to the invention.[78]

[75] DPMA, https://depatisnet.dpma.de/DepatisNet/depatisnet?action=pdf&docid=US000005785631A&xxxfull=1, accessed on 29.5.2025.

[77] EPO, https://register.epo.org/application?documentId=E6GX3EAR5431DSU&number=EP18778844&lng=de&npl=false, accessed on 22.10.2024.

[78] DPMA, https://depatisnet.dpma.de/DepatisNet/depatisnet?action=pdf&docid=DE102017122295A1&xxxfull=1, accessed on 2.1.2025.

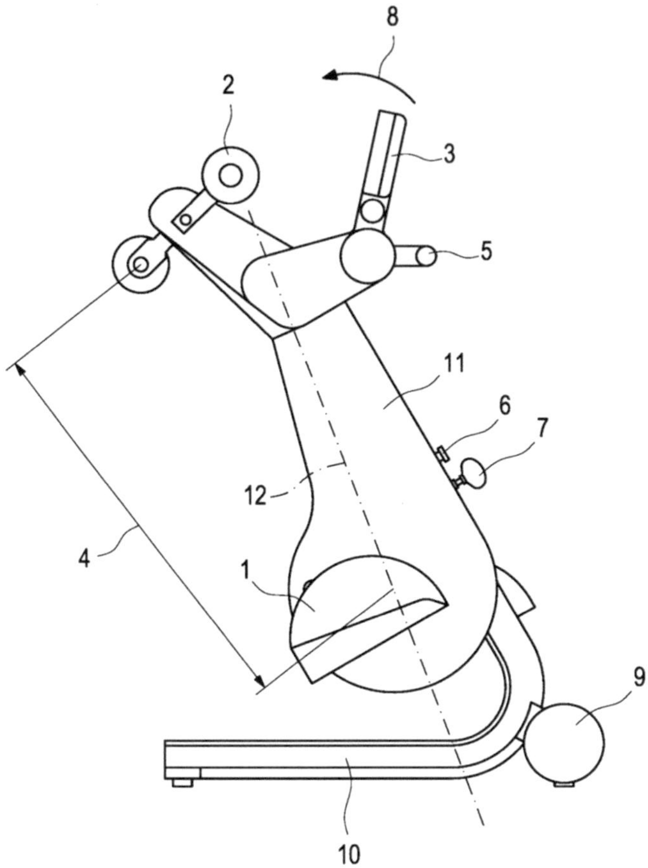

Fig. 7.34 Figure 1 of DE102017122295A1[79]

7.14 Example 10: Corner Connection for Sheet Metal Ducts

The applicant's own patent application EP 3964764 (EP 21193706.5) deals with sheet metal duct sections, which are used, for example, as ventilation systems. In the prior art, the individual sheet metal duct sections are screwed together, which is time-consuming.

The technical problem addressed by the application was therefore initially as follows:

"An object of the invention is therefore to provide a device that enables tight assembly of individual sheet metal duct sections. Furthermore, the device according to the invention

[79] DPMA, https://depatisnet.dpma.de/DepatisNet/depatisnet?action=pdf&docid=DE102017122295A1&xxxfull=1, accessed on 29.5.2025.

should reduce the production time required for assembling individual sheet metal duct sections."[80]

The main claim read as follows:

1. *"Connecting element (7) for releasably or non-releasably connecting a first and a second frame profile (4) to a first and a second sheet metal duct section (2), wherein the connecting element (7) is a hollow profile with a miter opening (6), which has a bend at the miter opening (6).*"[81]

Figure 7.35 shows the prior art described in the applicant's own application, in which the sheet metal duct sections 2 are fastened to the frame profiles 4 with connecting elements 5. The frame profiles 4 are connected by a screw connection 1 and 3.

Figure 7.36 shows the invention, in which the screw connections 1 and 3 of the prior art are replaced by profiles 7 with a miter cut.

Figure 7.37 shows the inventive connecting element 7 with the miter cut 6, which allows bending and thus results in a connecting element for a corner.

The patent office identified the prior art documents DE 29801851 U1 and US 5423576. Both documents also disclose corner connecting elements featuring miter cuts.

Figure 7.38 shows corner connections for sheet metal duct sections from prior art DE 29801851 U1, which feature a miter cut.[85]

Figure 7.39 shows sheet metal duct sections from prior art US 5423576, which are joined together using clip connections 30.[87]

Figure 7.40 shows a clip connection from prior art US 5423576, which is designed as a corner connecting element.

The applicant's own application EP21193706.5 included the following feature:

"By means of a special design of the miter cut 6, an additional fastening process by joining, folding, or welding can subsequently be added. In particular, the clip, the bent hollow profile 7, can be further stabilized by interrupting the miter cut on the rear side, at the duct-receiving leg. This leaves a strip of material on the other leg side, which can be folded over to one side after bending.

[80] EPO, https://register.epo.org/application?documentId=E6NSOLYT3392DSU&number=EP21193706&lng=de&npl=false, accessed on 6.12.2024.

[81] EPO, https://register.epo.org/application?documentId=E6NSOL200104DSU&number=EP21193706&lng=de&npl=false, accessed on 6.12.2024.

[85] DPMA, https://depatisnet.dpma.de/DepatisNet/depatisnet?action=pdf&docid=DE000029801851U1&xxxfull=1, accessed on 10.12.2024.

[87] DPMA, https://depatisnet.dpma.de/DepatisNet/depatisnet?action=pdf&docid=US000005423576A&xxxfull=1, accessed on 2.1.2025.

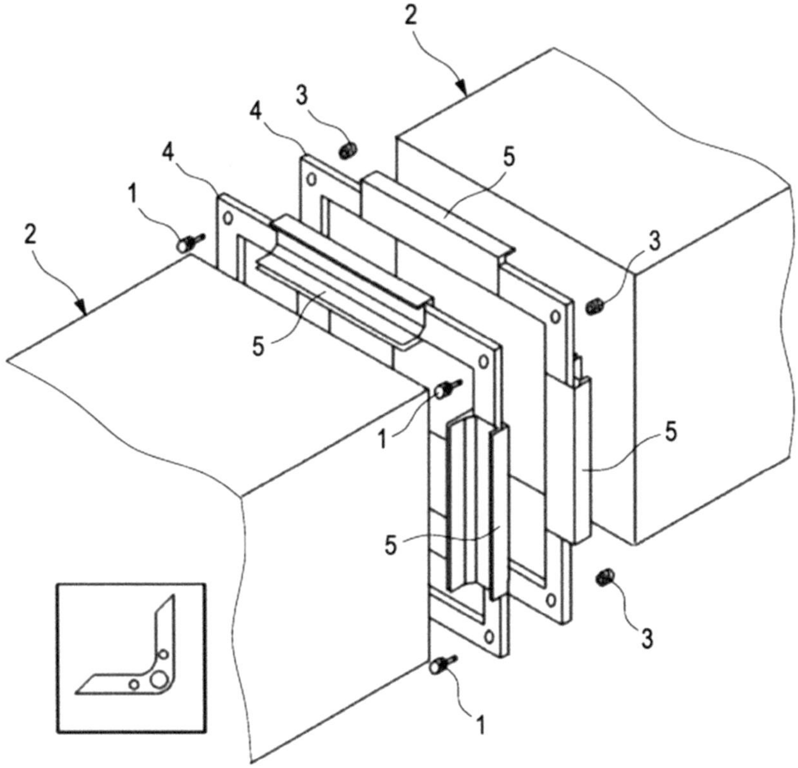

Fig. 2

Fig. 7.35 Figure 2 of EP3964764[82]

During further processing of the frame, for example when applying it to channel section 2, resistance spot welding or press fitting is performed. Such spot welding on the overlaps of the miter can increase the mechanical stability."[90]

From this, the feature could be derived that after folding, a mechanically stable corner connection element is produced by welding overlapping sections.

The new technical objective is therefore: To provide a corner connection element for joining sheet metal channel sections by clipping, which is designed to be particularly mechanically stable.

[82] DPMA, https://depatisnet.dpma.de/DepatisNet/depatisnet?action=pdf&docid=EP000003964764A1&xxxfull=1, accessed on 29.5.2025.

[90] EPA, https://register.epo.org/application?documentId=E6NSOLYT3392DSU&number=EP21193706&lng=de&npl=false, accessed on 12/31/2024.

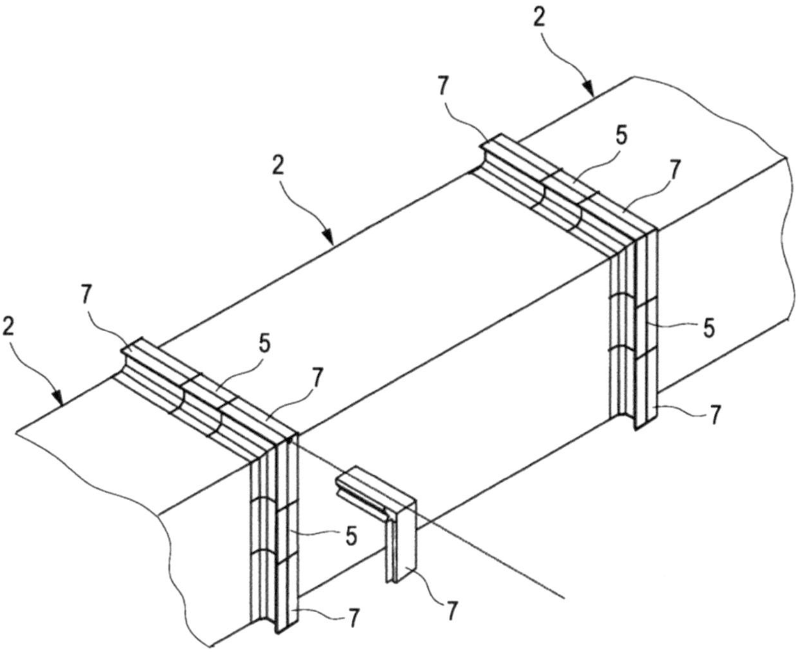

Fig. 8

Fig. 7.36 Figure 8 of EP3964764[83]

A new main claim could therefore read as follows:

1. Connecting element (7) for the detachable or non-detachable connection of a first and a second frame profile (4) to a first and a second sheet metal channel section (2), wherein the connecting element (7) is a hollow profile with a miter opening (6), which has a bend at the miter opening (6), wherein by folding at the miter opening (6), overlapping sections are joined by welding.

[83] DPMA, https://depatisnet.dpma.de/DepatisNet/depatisnet?action=pdf&docid=EP000003964764A1&xxxfull=1, accessed on 29.5.2025.

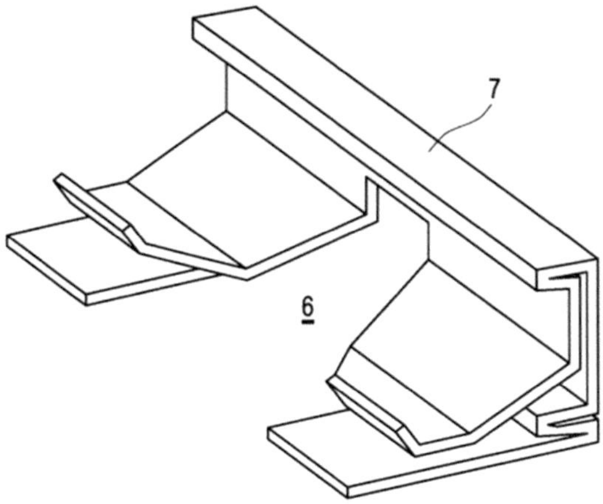

Fig. 5

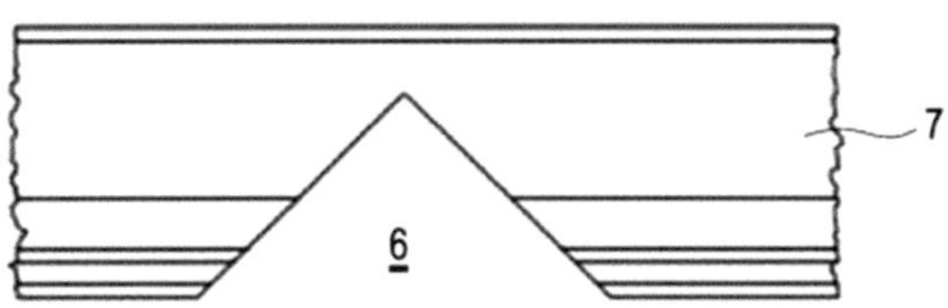

Fig. 6

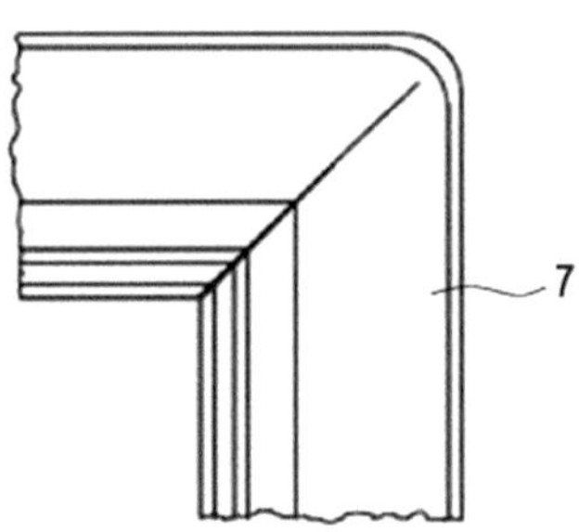

Fig. 7

Fig. 7.37 Figures 5 to 7 of EP3964764[84]

[84] DPMA, https://depatisnet.dpma.de/DepatisNet/depatisnet?action=pdf&docid=EP000003964764A1&xxxfull=1, accessed on 29.5.2025.

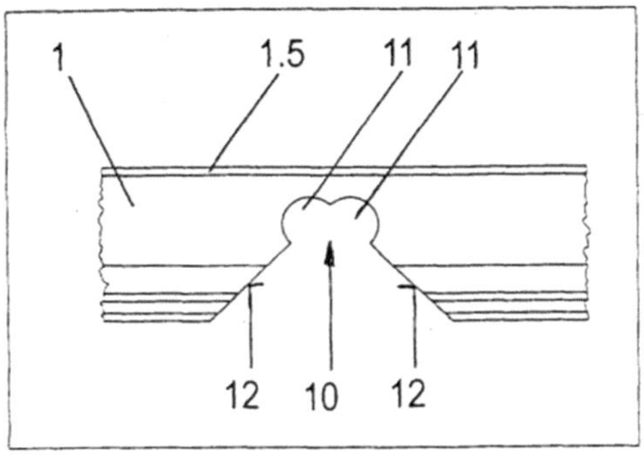

Fig. 1

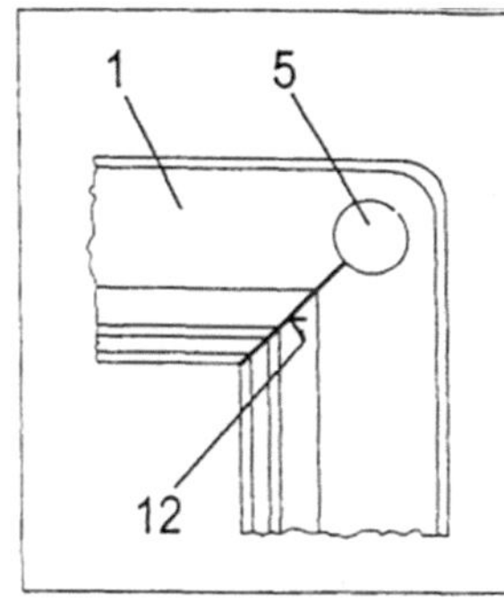

Fig. 2

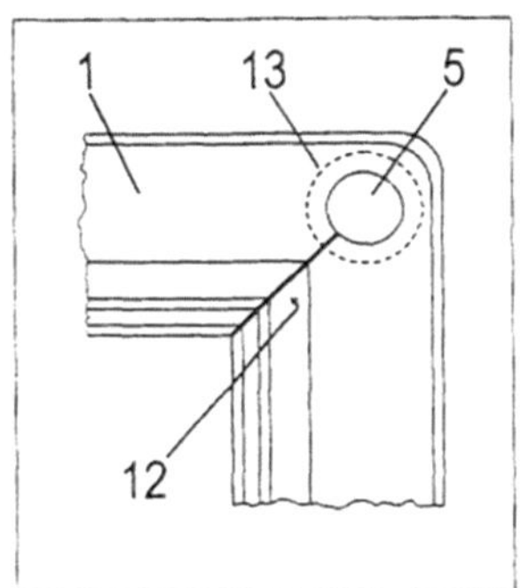

Fig. 3

Fig. 7.38 Figures 1 to 3 of DE29801851U1[86]

---

[86] DPMA, https://depatisnet.dpma.de/DepatisNet/depatisnet?action=pdf&docid=DE000029801851U1&xxxfull=1, accessed on 29.5.2025.

FIG. 1

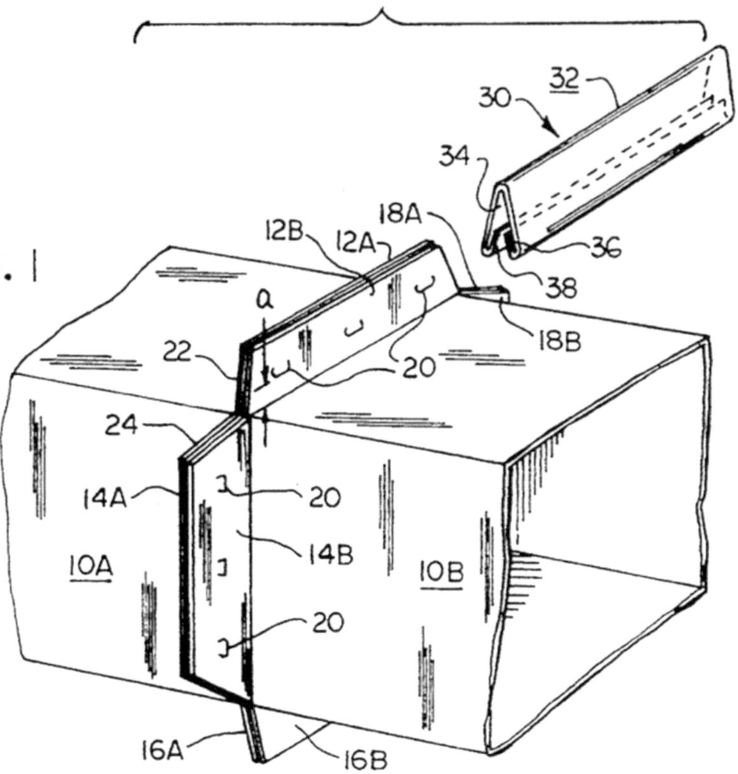

Fig. 7.39 Figure 1 of US5423576[88]

[88] DPMA, https://depatisnet.dpma.de/DepatisNet/depatisnet?action=pdf&docid=US000005423576A&xxxfull=1, accessed on 29.5.2025.

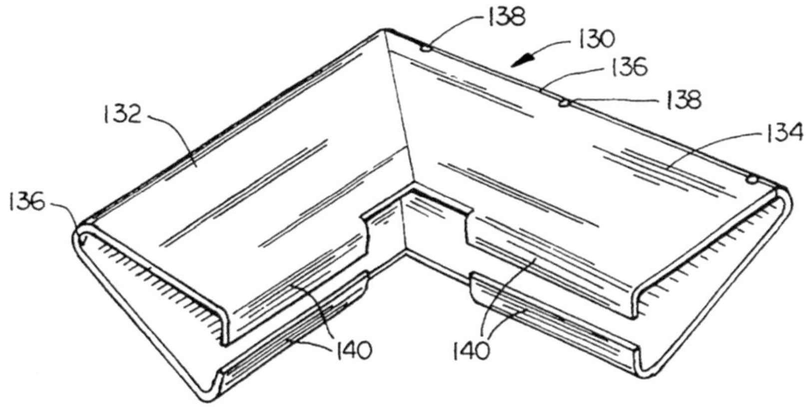

FIG. 9

Fig. 7.40 Figure 9 of US5423576[89]

[89] DPMA, https://depatisnet.dpma.de/DepatisNet/depatisnet?action=pdf&docid=US000005423576A&xxxfull=1, accessed on 29.5.2025.

MIX
Papier aus verantwortungsvollen Quellen
Paper from responsible sources
FSC® C105338

If you have any concerns about our products,
you can contact us on
ProductSafety@springernature.com

In case Publisher is established outside the EU,
the EU authorized representative is:
Springer Nature Customer Service Center GmbH
Europaplatz 3, 69115 Heidelberg, Germany

Printed by Libri Plureos GmbH
in Hamburg, Germany